CLEANING AND VALUE

sidestonepress

CLEANING AND VALUE

INTERDISCIPLINARY INVESTIGATIONS

edited by
Isabel Bredenbröker, Christina Hanzen & Felix Kotzur

Published by Sidestone Press, Leiden
www.sidestone.com

Lay-out & cover design: Sidestone Press
Photograph cover: Robert Schittko

ISBN 978-90-8890-921-4 (softcover)
ISBN 978-90-8890-922-1 (hardcover)
ISBN 978-90-8890-923-8 (PDF e-book)

Contents

Contributors

Mia Bencun, born in Lagos, Nigeria, lives and works in Offenbach where she obtained her diploma in Fine Arts at the Hochschule für Gestaltung. She received a scholarship from the Johannes-Mosbach-Stiftung in 2016, second place in the German-French Contemporary Art Rotary Art Prize 2016 and first place at the „Mit Kunst für die Kunst“ award in 2017 from Lions Club Darmstadt, the city of Darmstadt and Kunsthalle Darmstadt. Recent exhibitions include: Have you ever seen a demon? – Basis Projektraum, Frankfurt am Main, 2019; Colourful Chaos – by Fluxkunst, at Laundry Artspace, in Brisbane, Australia, 2018; body/tech. – at Museum für Angewandte Kunst, in Frankfurt am Main, 2018; Blockadia*Tiefsee 3 – at Kunstverein Freiburg, Freiburg, 2017.

Isabel Bredenbröker is an anthropology Ph.D. candidate at Goethe University Frankfurt where she was a staff member of the Value and Equivalence Research Training Group (2017-2019). She is also an Honorary Research Assistant at UCL Department of Anthropology. Isabel undertook her field research in the Ghanaian Volta Region. Her Ph.D. examines the materiality of death in Peki, Ghana. She has contributed to an anthropological sound exhibition on cleaning at the Bauhaus Dessau in 2015. Isabel completed an MA in Material and Visual Culture at the Department of Anthropology at UCL London (2014). She also holds an M.A. and a B.A. in Comparative Literature from FU Berlin (2013, 2009). Together with Philipp Bergmann, she produced the ethnographic short film ‚Now I Am Dead‘ (2019). She is also part of the a-disciplinary research platform ‚K‘ (www.khybrid.com) and works as an independent translator.

Georg Cyrus studied prehistoric archaeology at Free University Berlin. He achieved his B.A. in 2012 and his M.A. in 2016. In 2014, he took the opportunity to study at Boğazıcı University in Istanbul in the History Department. His special interest in dirt and cleaning behavior initially began during his early M.A. studies with a focus on contemporary archaeology, especially during the excavation of the forced labor camp in Tempelhof, Berlin. During excavations at the neolithic/chalcolithic site Kamiltepe, Azerbaijan, he began to think about dirt and cleaning behavior in a prehistoric context. After the participation in a project affiliated at the Tehran Department of the DAI he started his PhD at the Free University Berlin under the supervision of Prof. Dr. Susan Pollock in 2017. The working title is “Appropriation and conversion of monumental buildings at the example of Squatter occupations in the Iron Age of Southwest Asia”.

Jeanine Dağyeli studied Central Asian Studies and Islamic Studies at Humboldt University Berlin. After finishing her Magister Artium on healers and folk medicine in Kyrgyzstan, she worked as a doctoral research fellow at Martin- Luther University Halle-Wittenberg.

In 2008, she earned her Ph.D. on crafts-related moral codices of conduct. Afterwards Jeanine Dagyeli has worked as research coordinator for Gerda-Henkel Foundation at Al-Biruni Institute of Oriental Studies Tashkent (2009-2011), as post-doctoral fellow at Berlin Graduate School Muslim Cultures and Societies (2012-2013) and at Leibniz-Zentrum Moderner Orient (ZMO), Berlin (2014-2018). Since 2019 she is assistant professor at the School of Humanities and Social Sciences at Nazarbayev University in Astana, Kazakhstan.

Kerstin Gottschalk lives and works in Berlin and Frankfurt am Main. She completed her studies in painting at the Kunsthochschule Berlin Weißensee as a master student of Katharina Grosse in 2007. Since 2005, she has had several individual exhibitions and was featured in different group shows. In 2014, she received the Villa-Serpentara-Stipendium from the Akademie der Künste, Berlin. Since 2009, she has regularly taught at art academies. Since February 2017, she holds the chair for painting and graphic art at the Instiut für Kunstpädagogik, Goethe-Universität Frankfurt am Main.

Hans Peter Hahn is professor of Anthropology with a research focus on Africa at Goethe University, Frankfurt am Main. His research interests are material culture, consumption and the impact of globalization on non-western societies. He has edited a book on "Ethnologie und Weltkulturenmuseum" (Vergangenheitsverlag, Berlin 2017), focussing on the history of museums with ethnographic collections and is principal investigator in a research programme on virtual interfaces and museum collections (2017-2020). Until recently, he represented the research training group "Value and Equivalency" at Goethe-University, where he participated in the organization of several exhibitions. Other ongoing research initiatives are linked with polysemic approaches to material culture studies. Recent publications include an edited volumes entitled "Obstinacy of Things" (Neofelis 2015) and "Things as a challenge“ (transcript 2018).

Christina Hanzen studied Classical Archaeology, History and German Literature in Gießen and Münster. With her Master thesis she submitted a case study about the Charonian stairs, an architectural feature of the Greek theatre. She was a Ph.D candidate and staff member at the Research Training Group ‚Value and Equivalence‘, Goethe University Frankfurt. In her thesis she studies the displays of emotion in Greek Culture, focussing on their societal value and the negotiation of emotion in society.

Ainsley Hawthorn holds a Ph.D. in Near Eastern Languages and Civilizations from Yale University and has been a fellow of Distant Worlds, the graduate school for ancient studies at the Ludwig Maximilian University of Munich, and the Advanced Seminar in the Humanities, a collaboration between Ca' Foscari University of Venice and Venice International University. Her research interests include Mesopotamian literature and religion, Sumerian and Akkadian language, Middle Eastern dance, and the cultural history of the senses. She is co-editor of Distant Impressions, the first volume on the senses in the ancient Near East, which is forthcoming from Eisenbrauns academic press. Furthermore she is the Executive Director of the Community Foundation of Newfoundland and Labrador.

Andreas Koch is an artist, graphic designer and editor of the magazine 'von hundert'. He was born in Stuttgart in 1970 and has been living in Berlin since 1992, where he studied fine arts at UdK with Dieter Appelt and Christiane Möbius until 1998. He was awarded with the 'Berliner Arbeitsstipendium des Senats' twice (2000, 2011), as well as with grants by the Kunstfonds Bonn (2008) and Kunststiftung Baden-Württemberg. He has exhibited widely in- and outside of Berlin. Haus am Lützowplatz presented an overview of his works in 2015. Since 2006, he is an editor and author of the art magazine 'von hundert'. The magazine, which offers state-of-the-art critique of the Berlin art scene, was founded together with Kito Nedo and is currently co-managed with Barbara Buchmaier. (www.vonhundert.de)

Nikolaus Kockel studies at the Hochschule für Gestaltung Offenbach. Recent exhibition contributions include: 2019 "Pfad der Zerstörung" Le Bureau, Frankfurt. 2019 "Le Grand Trampolage" The Mythical Institution. 2019 "Mappenscout 24", Kunsthalle, Offenbach. 2018 "Comment ai je pu me perdre?" Goethe Institut, Paris. 2018 "Garten/kein Garten" Werkbundforum, Frankfurt. 2017 Performance "Cleaning and Value" Goethe-Universität, Frankfurt. 2017 "Vom Walde lernen" Station, Offenbach. 2016 und 2017 "But Heroes often fail" Hildesheim, Hannover, Frankfurt, Casablanca. 2016 "Future Nomads" CLB, Berlin. 2016 "Learning from Las Vegas.Reloaded" Streitfeld, München.

Ann Olga Koloski-Ostrow has done fieldwork at Herculaneum, Pompeii, in Jordan and Tunisia, as well as survey work in Croatia, Cyprus, Egypt, Israel, Lebanon, Libya, and Turkey. She works on Roman daily life, urban infrastructure, plumbing and hydraulics, baths and bathing practices, toilets and sanitation. She is Head of the Division of the Humanities, Kevy and Hortense Kaiserman Endowed Chair in the Humanities, Chair of the Department of Classical Studies, and Co-Director of a departmental graduate M.A. program Brandeis University. She holds an M.A. in Latin and Greek language and literature and a Ph.D. in Classical Art and Archaeology from University of Michigan. Publications include: "The Archaeology of Sanitation in Roman Italy: Toilets, Sewers, and Water Systems" (2018); "Roman Toilets: Their Archaeology and Cultural History" (co-edited with Jansen and Moormann, 2011).

Catrin Kost studied Sinology, Chinese Art and Archaeology as well as European Prehistory in Heidelberg, Munich, Shanghai and Paris. After receiving her PhD from Munich University, she held posts at the British Museum London, Heidelberg University and the Graduate School „Distant Worlds" (Munich University). Currently she is affiliated to the Department of Archaeology and Heritage Studies – School of Culture and Society at Aarhus University / Denmark. Dr. Kost's research focuses on ancient China, especially from the times of the Warring States (475-221 BCE) to the Han dynasty (206 BCE-9 CE; 25-220 CE) and addresses questions of mobile-settled interaction, cultural exchange and receptivity as well as every-day material culture. The paper addresses an aspect of her habilitation treatise: Von „Schmutzfinken" und „Saubermännern" – Körperpflege und Körperlichkeit als Spiegel sozialen Wandels im alten China (1045 v. Chr. – 220 n. Chr.). Publications include: "The Practice of Imagery in the Northern Chinese Steppe (5th-1st centuries BCE)" (Bonn 2014).

Felix Kotzur studied Roman Provincial Archaeology, Classical Archaeology and Archaeometry in Frankfurt am Main, Munich and Leicester. He received his Magister Artium degree with a thesis on two excavated ditches from the Roman period in 2015. From 2016 until 2019, Felix was a Ph.D. candiate and staff member at the Research Training Group 'Value and Equivalence', Goethe University Frankfurt. His Ph.D. research explores intercultural contact between the Roman Empire and its neighbours, especially the comparison of grave inventories and rituals. Since 2019, he works at the Deutsche Limeskommission, which is responsible for the structural organisation and publication of insights on the world heritage site 'Upper German-Raetian Limes'.

Robinson Peter Krämer studied Classical Archaeology, Ancient History and Prehistory at the University of Bonn. He received a PhD Scholarship from the German Academic Exchange Service and was a Ph.D. student in the DFG Research Training Group 1878 – Archaeology of Premodern Economies in Cologne and Bonn. Krämer received his Ph.D. in December of 2016 for a thesis on Etruscan Sanctuaries (8th-5th centuries BCE) and their functions as economic spaces.Since April of 2017, he is a Postdoctoral Research Associate at the Heinrich Schliemann-Institut für Altertumswissenschaften, Department of Classical Archaeology at the University of Rostock. Krämer's research focuses on Central Italy from the 10th-4th centuries BCE, contextual analysis of pottery and the reconstruction of social and political developments with archaeological sources.

Olli Lagerspetz is a professor of Philosophy at Åbo Akademi University. In 1992 to 1998, Lagerspetz was enrolled as Lecturer of philosophy at the University of Wales Swansea. Lagerspetz is the author of "Trust: The Tacit Demand" (Kluwer: Dordrecht, 1998) and "Smuts: En bok om världen, vårt hem" (Symposion: Stockholm/Stehag, 2006) [Dirt: A Book On the World, Our Home].Lagerspetz's main research interests include philosophical anthropology, philosophical psychology, moral philosophy and the philosophy of mind. Theoretically, his work is influenced by Wittgenstein and post-Wittgensteinian philosophers. A consistent theme in his work is the relation between theoretical accounts and everyday practices, and the need to develop alternatives to reductionist tendencies.

Anna Langgartner studied classical archaeology and art history at Goethe University Frankfurt and obtained her M. A. in 2018. During her studies, she worked at excavation projects in Priene (Turkey) and Olbia (Ukraine) for several years. Since 2017, she works as a guide for Liebieghaus, Frankfurt and currently holds a research training position at the Roman Fort Saalburg museum, Bad Homburg. Anna's Ph.D. investigates religion and cults at roman forts alongside the Upper Germanic-Rhaetian Limes.

Chrischa Venus Oswald is an artist and poet interested in exploring the human condition and relationships of all kinds. She finished her Fine Arts studies at the University of Art and Design in Linz with an honours diploma in 2011. In 2007 she was awarded the Diesel New Art Award Austria for Photography. Chrischa's work was exhibited and screened internationally in group- and solo shows and is included in private collections such as the video collection of Manuel de Santaren. She works predominantly with photography and video. Much of her work is based on performative or narrative concepts tied to personal experiences, myth(ology), specific knowledge or stories. She lives in Lisbon, Portugal. Her work can be found online at www.chrischa-oswald.com

Wagehe Raufi studies at the Hochschule für Gestaltung in Offenbach. Between 2018 and 2019 she spent a semester abroad at Central Academy of Fine Arts in Beijing, CN. Recent exhibition contributions include: Videonale.16, Bonn (2017), blockadia•tiefsee, Kunstverein Freiburg (2017); Exhibition of Residence Artists, Masterhouse Gropius Dessau (2017); Einsteins Spürnase, Leopoldina, Halle (Saale) (2018); body/tech, Museum Angewandte Kunst, Frankfurt (2018); 88. Herbstausstellung, Kunstverein Hannover (2018); And is is Us, Frankfurter Kunstverein (2019); Shift, Frappant Gallery, Hamburg (2019); L'Arbre du Soleil, Mountains Gallery, Berlin (2019); This is So Me, Opelvillen Labor, Rüsselsheim (2020).

Robert Schittko is currently working on his diploma thesis about the change of photographic techniques in applied art. His most recent exhibitions are the Athens Photo Festival 2019, the exhibition "Fail early and often" within the SAP Art Collection and other group exhibitions in Berlin, Frankfurt and Paris. He was awarded with the Prize of the Deutsche Börse Photography Foundation: "I have no idea how i could get that lost – Aokigahara", shown in Frankfurt am Main.

Mareike Späth's work focuses on historiography and practices of the past in the present. She is particularly interested in entangled histories and heritages of Africa and Europe, their (non-)representation in public places and public history, as well as unofficial and alternative narratives. Having spent some years researching and lecturing at the Department of Anthropology and African Studies at Johannes Gutenberg-University Mainz and the Department of Social and Cultural Anthropology at Goethe-University Frankfurt, she now works as a researcher and curator at the ethnographic collection of the Hanover State Museum.

Ulrich Veit is a professor for Pre- and Protohistoric Archaeology in Leipzig. His main research interests are in the field of later European prehistory with special interests in social archaeology as well as in the historiography and theory of archaeology.
Recent publications are: "Donnerkeil – Opfermesser – Thränengefäß. Die archäologischen Objekte aus der Sammlung der Leipziger Apothekerfamilie Linck (1670-1807) im Naturalienkabinett Waldenburg (Sachsen) " (2014, ed. with M. Wöhrl) and „Der bandkeramische Siedlungsplatz Eythra in Sachsen. Studien zur Chronologie und Siedlungsentwicklung" (2016, ed. with H. Stäuble).

Sebile Yapici studied Turkology and Central Asian Studies in Berlin and Istanbul. She is currently writing her Ph.D. in Social Anthropology and lives and teaches in China. Her work focuses on Uzbek food culture and seeks to explore if and how the perception of the value of food and eating together changes in diaspora. She considers the consumption of one specific rice dish, to understand its impact on the cultivation of social relationships.

Editor's note

We would like to thank all participants of the workshop who have contributed a paper, artwork or text. We'd particularly like to thank those who have journeyed on with us and contributed to this publication which has been a time-consuming test of patience. This is our first academic publication as editors and we are happy to have learned a lot in the process, hopefully resulting in an interesting and diverse book that will engage readers in interdisciplinary thoughts on cleaning and value.

We would particularly like to acknowledge and thank the research training group Value and Equivalence at Goethe University Frankfurt and the DFG, which have provided us with the funding and infrastructure to take on such a project and invite scholars to participate. Our heartfelt thanks particularly include Professor Hans Peter Hahn who has supported this book project as well as Annabel Bokern who guided us along the way and gave us valuable advice on budget questions. We'd also like to thank Stefanie Samida for her feedback on our joint editorial introduction, Daniel Gottlieb for priceless proofreading and editorial feedback as well as our very patient and thorough student assistant Mareike Chudaska for editing and proofing citations and bibliographies.

Figure 1. Participants of the workshop (Photo: Research Training Group 'Value and Equivalence').

Preface

Hans P. Hahn

Is it possible that 'dust from the Louvre' can be regarded as a special and authentic 'object' of value? Cologne-based collector Wolfgang Stöcker founded an 'international archive of cultural dust', in which he keeps dust samples from various important places, among them valuable dust from Europe's most important museum – the Louvre in Paris. If we follow the collector's opinion, the previously worthless dirt – removed from its original location, carefully stored and labelled with a precise description – has become a document of cultural history. Stöcker's samples originate from places where culture is staged and publicly experienced. Although the contribution that dirt makes to the constitution of such special places might be contested, its cultural value can hardly be denied.

Behind each object in his collection stands an act that is familiar to everyone as universal routine – the act of dusting. Regardless of whether he finds himself in the Louvre or in St. Peter's Basilica in Rome, Stöcker wipes the frames of pictures or the backs of pews and collects dust that many other visitors had probably not payed any attention to at all. Wiping, cleaning and separating clean and dirty things are undoubtedly some of the most fundamental, culturally defined ways of assigning certain material items a clear place in society and within the associated order of values. Normally, these processes yield an unambiguous allocation of value and worthlessness.

But, is it not also true that even within such separated matter – within dirt – there remains a trace of the valuable? The importance of purification as an economic practice is a product of blurred definitions. In particular, one should think of archaeology, within which some deposits, remnants or items of dirt have a high information value with regard to a particular culture or monument. It is the aim of this book to highlight the hidden connotations of cleaning, the value of dirt and the dynamics of the concept of value as a whole.

In this sense, this edited volume represents a substantial contribution to the topic of the Research Training Group "Value and Equivalent" (RTG 1576). The relevance of this topic is not just restricted to the problematic dichotomy of "clean and dirty" or "valuable and worthless". It also responds to an inherent and extraordinary transdisciplinary scope. Hardly any discipline of the humanities can escape the significance of "purification" as a culturally defined and culturally productive activity. Purification – at least in the non-metaphorical sense of Stöcker's collection – inevitably leads to the production of dirt.

I would like to congratulate the editors on their choice of topic and the compilation of this anthology. With a fresh approach and a clear theme that has hardly been reflected on in the humanities to this date, they have succeeded in engaging authors from various disciplines and compiled a collection of original contributions. It should be emphasized that the authors who have contributed to this volume do not restrict their engagement to popular theoretical references, such as for example Mary Douglas's work on dirt and cleanliness. Instead, the contributions present original case studies that lead to very specific, sometimes surprising interpretations of the value of cleaning and the value of apparent dirt.

The Research Training Group has gladly supported the publication of this volume. In addition to the obvious relevance of the topic, which had already been established during the preparation of the workshop on which this book is based, the bigger aim of this funding scheme is to provide training for young researchers in good scientific practice. Of course, this also includes the publication of an anthology. Editorial work in particular demands an extraordinary amount of commitment from those involved, since all contributions had to be supervised, accepted and edited as well as subjected to a review process. Yet, the practical experience of taking editorial responsibilities is also an important step on the way to a future academic career. The editors mastered all the necessary tasks and the present book is a valuable proof of the efficiency of the Research Training Group. On behalf of the group, I would therefore like to thank the editors and the contributors for their work.

Frankfurt / Main, December 2019

Paper Abstracts

Isabel Bredenbröker presents an ethnographic case study from a Ghanaian Ewe town in the Volta Region. Here, death motivates a range of practices and activities that deal with the production of a 'good death'. *The Last Bath: Cleaning Practices and the Production of 'Good Death' in an Ewe Town* looks at the political relevance of cleaning practices related to death and everyday cleaning practices. By looking at the materiality of bodies dead and alive, at the material qualities of things used for cleaning as well as political structures in the community, the paper describes how symbolic and political values are negotiated through cleaning.

Georg Cyrus examines possible traces of cleaning at the archaeological site Fistikli Höyük in his paper *The Cultural Aspect of Cleaning in Archaeology – A Case Study from the Late Neolithic Site of Fistikli Höyük*. The archaeological record discussed comprises organic remains and lithic refuse. The paper tests whether it is possible to know what the Neolithic people considered as clean and unclean given the archaeological finds and the elusiveness of categories such as clean or dirty in cultural contexts.

Jeanine Dağyeli's paper *Why is Death Defiling? Considering Death-related Pollution and Cleaning in Central Asia* looks at corpse washers in Uzbekistan and Kyrgyzstan. Washing here is usually performed by family members but some communities use professional corpse washers for fear of pollution. Dağyeli shows how cleaning in this context affects both the material as well as the spiritual world.

Ainsley Hawthorn's paper *Your Clothes Should Be Clean! Your Head Should Be Washed! Body Cleaning and Social Inclusion in the Epic of Gilgamesh* reads the friendship between Enkidu and Gilgamesh in the Gilgamesh Epos as an example of social valuation through body cleaning in Ancient Mesopotamia. Enkidu grows from beast to human by means of initiation to body cleaning. In contrast, the young king Gilgamesh takes ritual baths and his awareness of status is much higher. Hawthorn shows how the epos frames body cleaning as a marker for social distinction.

Ann Olga Koloski Ostrow's paper *CARPE Dirt, Disease and Detritus: Roman Sanitation and its Value System* reviews understandings of hygiene in Roman antiquity. She investigates the construction and purpose of sanitary facilities, water supply and drainage systems by discussing texts as well as archaeological finds from Pompeian sites and the city of Rome.

Catrin Kost's paper *A Matter of Representation – Personal Hygiene in Eastern Zhou-dynasty China (771-256)* looks at bronze vessels in ritual and funerary contexts and investigates the socio-political function of the vessels which were explicitly labelled as washing tools. The paper looks at twenty-seven bronze vessel finds and examines their emergence, circulation and usage.

Robinson Krämer's paper *The Cleansing of a Political System: Obliterations, Burials and 'Reuse' of Palaces and Seats of Power in Central Italy (Seventh-Fifth centuries BCE)* explores architectural transformations in the wake of political changes based on archaeological finds from Central Italy on Etruscan and Latin territories, covering the time between the seventh and fifth centuries BCE. The paper considers the practical use and changes made to architectural seats of power that occurred here when a power-centralising political system such as monarchy, tyranny or autocracy was overthrown by more democratic movements.

Olli Lagerspetz's paper *How is Dirt Possible? On the Philosophy of Dirt, Cleanliness and Refuse* tracks the complexity of dirt and cleanliness via a contrasting discussion of philosophical and modern scientific approaches. He discusses ideas which he links to the terms scientific realism and culturalism. Against these two, Lagerspetz bases his own approach on Aristotle and argues that caring for objects generates consciousness of what is polluted and clean in the sense of a perfect, ideal state.

Mareike Späth's paper *Cleaning up the Past* examines attempts at eliminating unwanted historical narratives from public and collective memory through cleaning. Späth discusses the history of monuments, their alterations, destruction and afterlives in Madagascar and beyond. Pieces of monuments may survive as museum objects and their changed status may not necessarily alter memory or change the narrative of history.

Ulrich Veit's paper *Archaeology and Cleaning: Some Reflections on the Archaeological Process* examines the potential and boundaries of cleaning in the context of archaeological excavation practice and theory. Obviously cleaning is indispensable for every excavation process. Nevertheless, it is also inextricably linked to the destruction of contexts at excavation sites. Understanding how cleaning was practiced is difficult to show by means of archaeological evidence and yet archaeology deals with the remains of cultures, things that have been abandoned and forgotten. The paper discusses these paradoxa and links them to contemporary practices.

Sebile Yapici's paper *Shaking out the Tablecloth – Uzbek Hospitality and the Construction of Boundaries of Belonging* discusses hospitality and food in Uzbek culture. Cleaning and cooking are intrinsically intertwined in Uzbek households yet the work of washing up and cooking is gendered work, almost exclusively reserved for women. As strangers and friends play the role of guests and older women and men host, young women and new brides can earn their position as insiders and kin of the household by cleaning. The making of an inside and outside – kin and friends, strangers and new family members – is negotiated through hosting. On a larger level, so is the question of national belonging – being Uzbek – and of performing Uzbek culture in the North American diaspora.

Section One

Editorial

We have never been Clean – Towards an Interdisciplinary Discourse about Cleaning and Value

Isabel Bredenbröker, Christina Hanzen, Felix Kotzur

The contributions to this interdisciplinary volume attempt to combine thinking about cleaning with thinking about value. At first glance, the way in which cleaning relates to the changing value status of material cultures, environments, bodies or persons seems obvious. Cleaning may be understood and practised as maintenance work, renewing or generating value through a constant process of care. But cleaning may also be an uninvited, damaging or violent change in the status of things, for example when it leads to increases in real estate and rental market prices as can be noted when looking at processes of gentrification in contemporary cities across the globe. Cleaning may also be the ideological or virtual creation of categories, for example of national belonging, race and culture, social status, gender etc. Categories in science and academic research have furthermore come to typify the project of modernity, which relies on the neat organisation of objects, observations and conditions into a clean-cut order. As the archaeological contributions to this volume especially show, however, assessments of clean and dirty states – as well as practices to work around them – have existed in many different forms throughout history and cultures.

Cleaning may represent the execution of power and the suppression of forces trying to overthrow it; it thus has both political relevance and a practical function. These are the grounds on which thinking about cleaning and value becomes truly interesting. For all of these different scenarios, the difficulty lies in distinguishing the immaterial dimension of cleanliness as an evaluation or symbolic property in relation to specific material properties and conditions. The tension between moral or socially constructed values and ideas around 'good' cleaning and the reflection of such ideas in the material world means that the terrain in which values are formulated is always subject to dramatic or surprising shifts in perspective (fig. 1).

This introduction seeks to give an overview of relevant theoretical positions towards cleaning and to connect them to the field of value theory. We are considering and discussing positions from anthropology and archaeology as well as research approaches from the natural sciences, art and related social sciences. And while the term 'value'

in: Bredenbröker, I., Hanzen, C. & Kotzur, F (eds) 2020: *Cleaning and Value. Interdisciplinary Investigations*, Sidestone Press (Leiden), pp. 23-38.

Figure 1. Word cloud Cleaning and Value. This cloud illustrates in a term-quantity-relation the themed table discussions during the workshop; the bigger a word is, the more often it was mentioned by the debaters (Designed by Sara Kessler).

covers a broad range of definitions and applications, it is nevertheless apparent that it has a stronger presence in these different discourses than the term cleaning does. We would like to therefore start by investigating the potential of cleaning as a term and practice that is, in the words of Claude Lévi-Strauss, 'good to think with'.

The strong ontological Status of Dirt and the tedious Work of Cleaning

Despite the praxeological turn, the practice of cleaning – formulated as a verb, implying activity – is not a key term in canonical texts from anthropology, archaeology or affiliated disciplines of the humanities and social sciences. Instead of a discussion about activities or practices, the identification of states has received more attention. Dirtiness and purity mark a dynamic field of tension. Linguistic differences, particularly in English, are good markers to think about the nature of such tensions as well as about the social processes within which these conditions are incorporated. In literature, there seems to be a focus on dirty states and dirt rather than on the somewhat ephemeral non-materiality of the clean state. In anthropology, the notions of dirt and pollution have been discussed since Mary Douglas' (1969) key text *Purity and Danger*. This text is, strictly speaking, already impure in a valuable way, since it assembles its sources and creates its argument through various references from history, religious studies as well as ethnographic material. The central 'matter' of Mary Douglas' text is not so much purity, which may be considered a material property, but dirt as a marker of social boundaries and limits to a symbolic order. Douglas' approach remains an invaluable resource to think about how dirt acts as a marker of a tangible state of disorder, identified as 'matter out of place' (see Duschinsky, Schnall & Weiss, 2016). Dirtiness, interestingly, does not seem to be reduced to being a property or a condition of materials and things. It may take over and become the defining signifier, not just "dirty", but "dirt", changing the ontological status of a thing (at least in English). When trying to apply this to cleanliness or purity, "clean", taken as a noun, does not seem to apply to the matter in the same way or with the same force. Things, materials and bodies may be in clean conditions, but this ultimately never fully defines their status of being. You can find "dirt" lying around in the corner of a room, but the shiny floor cannot be described as "the clean" or "the purity" one walks on. The clean state is a precarious condition without nominalisation while the dirty state is taken to be a much more fundamental one, defining the way matter is perceived: as dirt. Since dirt in this strong ontological meaning has materiality, it is easier to think about, identify, challenge, act upon and refer to it, which might explain the prominent role that it plays in Douglas' text.

The activity of cleaning could therefore be considered a truly transformative practice that may undo the ontological status of matter as 'dirt' and return it to a precarious, yet possibly more desirable state of being clean. Words like filtration, extraction or sterilisation, which are usually used in natural sciences are indicative of this. They name processes of transformation at the end of which a substance, object or living being is found to be in a state of purity or cleanliness. In such a clean state, a thing, material or body has restored to it the potential of having its own, proper ontological definition. The assemblage of fibres, dust and other particles that are usually identified as 'household' dirt when lying in the corner of a room may only be regarded as something other than that in contexts where it would be taken to be 'in place'. These contexts may, for example, be in the laboratory as a sample, on the street rather than in a room or in other orders that make sense of it. An object – for example a plate smeared with food – is less likely to be categorised as dirt after having been used for eating. Remains of the meal can be washed off, restoring the open state of the plate as a useful thing, at hand for another person to use. In circumstances, however, where the plate does not represent a functional object for eating – let's say if it is decorated in an undesirable way, it may risk becoming dirt or at least being categorised as polluted in a way that is hard to clean up or alter. The likeliness of such a plate being unwanted is high. What happens now, be it re-contextualisation, physical alteration of the object or its removal, is a social practice that reveals the ways in which orders are made, or unmade, as Douglas frames it. Continuing Douglas' thoughts, the practice of cleaning and not simply identifying clean 'states' stands at the centre of this volume.

The work of Purification and Modernity

Bruno Latour (1994) has offered a reading of modernity as a non-state which consists of two opposing realms of practices, 'the work of purification' and the creation of hybrids:

> *"As I have said, we have to trace both the modern and the non-modern dimension, we have to deploy the latitude and longitude that will allow us to draw maps adapted both to the work of mediation and to the work of purification"*, p.77.

Latour (1994) identifies the work of purification here as any process which attempts to translate hybrid things into their components 'by conceiving every hybrid as a mixture of two pure forms' (p.78). Purification may be understood as a cleaning process that works in such a way by dissecting and breaking things up. It also seems to have a particular, ideal outcome in mind. According to the modern process of 'cleaning' social spheres, technical innovation and other tools of modernity such as scientific categories and methods aim at creating a new, purified way of understanding and relating to the world. Modernity claims to separate itself from past strictures, such as for example religious beliefs as well as the non-modern, non-European practices related to it. The ideal modern outcome is the production of a secular state (Chakrabarty et al., 2015a, 2015b) At the same time, this modern ideal has been proven faulty and idealistic. Nacim Ghanbari and Marcus Hahn (2013), for example, have used Latour's thoughts on 'the work of purification' to assemble perspectives that frame processes of modernity in the field of tension between (modern) media and religion, highlighting the hybrid forms that exist here, which do not produce a secular (eg. modern and clean) state, but something new. In the field of technology and religion alone, there are multiple texts and studies which highlight the hybrid forms and practices which reproduce 'a-modern' practices of using modern techniques (Meyer, 1999; Meyer, 2015; Stolow, 2013).

Latour describes modernity as a cosmology which, although proclaiming to provide a definitive separation of nature and culture, has to abide by the same rules as other cosmologies do: the separation of things through the 'work of purification' and the creation of hybrids through mediation or translation. Latour identifies processes of cleaning in interaction with the material world as well as in social practices and interactions, in the creation of categories and the evaluation of things, people and states according to such categories. Purification has become a leading modern ideal and yet deemed to be permanently corrupted by hybridisation and hybrids. Latour's ideas are set on the level of a "grand" theory of the social and aim at commenting on the project of modernity as a "grand" project seem to match the linguistic specificities described previously. Dirt, a hybrid par excellence, has a stronger ontological position which is conveyed through its nominal form. The state of cleanliness, for Latour, is synonymous with the ideal of modernity. It is always about to collapse and be made impure due to its weaker ontological status. Purity in its self-referential form cannot be found in the world, it can only be attributed to its parts.

Natural sciences translate cleaning processes into a rather neutral scientific sphere, where it is unusual to distinguish between clean and dirt, between the "good" and the "bad" molecule. Furthermore, natural sciences draw seemingly arbitrarily lines in their approaches towards defining cleanliness. Above all, the artificially created conditions of a laboratory do not match the general idea of a "natural" starting point. Yet, the condition of nature may at times be considered both clean and desirable. The abstract of a book about protein purification applications illustrates the flexibility of cleaning definitions. Here, an intentional scientific process of purification is described as a 'strategy that can produce a highly pure single protein from a crude mixture of proteins, carbohydrates, lipids, and cell debris is a work of art to be admired' (Roe, 2001). The isolation of a pure, single protein is the desired result of described scientific purification process. The scientists "clean away" all other components that initially seem to be naturally attached elements. Yet, they are not part of the intended, clean product. The definitions of clean and pure are flexible in the realm of laboratory and scientific processes. Nature appears just as unstable and prone to shifts in definitions as social concepts of ordering.

Figure 2. Cleaning the sea from oil spills (Doug Helton, NOAANOSORR Flickr).

In addition to Latour's argument in which science itself plays a significant part in the formation of the categories pure and impure, there are also instances in which science considers nature a clean environment and not just as a possible threat to modern, "clean" categories. Geert Potters (2015), for instance, describes marine pollution as a phenomenon occurring when a 'natural' habitat is contaminated with liquids, solids or any other state of matter, which cause an imbalance of compounds, showing a higher concentration than normal. In the case of oil or plastic particles in the sea, these are direct effects of human activity and is, as such, 'man-made' pollution (fig. 2).

In this context, the removal of certain particles from seawater, reducing it to its natural state, can be described as cleaning it from pollution. Human intervention therefore acts as a pollutant as well as a cleaning force, whereas, in this scientific understanding of nature, the original state of the marine lifeworld is considered a clean state.

Cleaning from an Interdisciplinary Perspective

This volume brings together interdisciplinary perspectives on the process of cleaning as an activity which may be the flagship of modernity but which carries within it the potential to turn into its very opposite. The volume does so by crossing the clean lines between disciplines and initiating a dialogue that includes different branches of anthropology and archaeology, as well as history, philology, philosophy, natural sciences and the arts. Although academic thought is also a product of modernity, anthropology gives voice to non-modern positions in the contemporary world, archaeology interprets social life that has long ceased to exist by its very non-modern material traces, history is itself devoted to social life well before "moderns" came along, art history traces different forms of artistic expression across time and space and so on. And while historical sources do not allow us to observe practices as they occur, they still carry evidence of their existence.

In anthropology, as already discussed at the beginning, studies that have been produced on issues of cleanliness, dirt and processes of cleaning are situated in the field of waste studies (Thompson, 1979), studies of re-use and recycling of materials and objects (Alexander & Reno, 2012; Norris, 2010), rituals of cleanliness and pollution as well as the touching points of holy and dirty states (Alley, 2002; Douglas, 1969) and studies of materials (Drazin & Küchler, 2015). Bruno Latour and Steven Woolgar's (1986) ethnographic account of the laboratory is also part of this tradition, making explicit use of ethnographic material which describes the production of a clean laboratory state that provides explicit categories of pureness for one of modernity's main tools: the natural sciences.

How far this branch of research has developed outside of anthropology since Thompson's initial publication can be noted in works like the *Encyclopaedia of Consumption and Waste* (Zimrig & Rathje, 2012). In these two volumes, experts in the field of refuse archaeology and environmental science compiled the actual state of the research and provided definitions of important terms. A recent example of archaeological research in this direction is the volume *Archaeology of Waste: Encounters with the Unwanted* (Sosna & Brunclíková, 2017). The editors stress that dealing with waste and similar topics is still unusual in archaeology despite being a promising field of research. This circumstance may be surprising, given that field archaeology deals with cleaning and waste on a daily basis. Common German excavation terms and actions like *die Fläche putzen* (cleaning the slate) or *Scherben waschen* (washing the shards) show this quite literally. But the fact that every excavation comes with the problems of cleaning the structures and the objects that are uncovered has not led to a discussion of this topic in the form of a thorough theoretical analysis. The archaeological contributions in this volume will show an emergent awareness of the concept of cleaning within the discipline. A few publications, however, have already grappled with dirt and waste in different contexts. They are mainly case studies, for example about domestic waste management (Felix Arnold, 2014) or on the question of dirt and depositions (Mike Parker Pearson, 2001). Furthermore, the short contribution from Ulrich Veit (2005/2006) and the investigation of waste as a source of archaeological knowledge by Dietmar Schmidt (2005) are worth mentioning. It is safe to assume that the concept of cleaning is still to be discovered by the major branches of archaeological thought and practice, but its value is already recognised.

Making Art with Rubbish and Bodily Fluids

In art, which, as Alfred Gell (1992) points out, is in itself a value system, the play with unusual or potentially polluting materials has traditionally been an artistic strategy to challenge existing orders. This has served to critically comment on them and address events or expressions of violence, such as through the use of bodily fluids, dust, refuse, trash, rubbish, "poor" materials or repulsive aesthetics. Naming only two examples, the Mexican artist Teresa Margolles vaporised water that was previously used to wash bodies in a morgue in Mexico City in an installation. The Spanish concept artist Santiago Sierra blew car exhaust in the interior of a synagogue in reference to the Holocaust. Art has also been treated as dirt in many cases where the value framework of artistic contextualisation and production was crossed by other practices of evaluation. Joseph Beuys' *Fettecke*, an installation that consisted of thickly applied grease in the corner of a museum, was completely removed by a cleaner who mistook it for dirt. This shows quite clearly how the cleaning up of an artwork arises due to a clash of valuations and related practices and contexts: that of the artwork and the museum environment, both of which are charged with ensuring the market price value of the work and the value of artistic production and of the uniqueness of the work. In contrast, stands the evaluation of grease as a substance out of place in the corner of a museum. A misunderstanding arises when the values of art come into conflict with the cleaner's more practical, socially embedded value of cleanliness. The cleaner, in the end, destroys the artwork and annihilates its materiality, hereby possibly and paradoxically even increasing the work's fame and value.

Cleaning and Value

The natural sciences endeavour to achieve all sorts of valuation: de-, re- and "trans-valuation". The predominant ambition is re-valuation, since cleaning does here seem to have a predominantly positive connotation. Either the positive, intended result lies in the re-establishment of an earlier, pre-polluted state like the sea without mineral oil or, in the maintenance of technical installations, such as when solar cells are kept free from desert dust. The destructive element that always accompanies processes of movement, separation or dissolution is terminologically neglected.

But cleaning is not always deemed beneficial when it comes to economic evaluations. An example is the Ganges in India, which is heavily polluted by industry and agriculture. The river is an important source for livelihood and prosperity, and as a deity, it is vital for local religious practices. Regarding its state of pollution and the threats that this poses to humans and the environment, cleaning the river seems beneficial for those who use it as a resource. The application of these cleaning practices is limited, however, by the financial reality of managing such a project. The title of the book *Cleaning up the Ganges. A cost-benefit analysis of the Ganga Action Plan* (Markandya & Murty, 2000) already reveals the difficulty for the public to arbitrate between the value of a clean river and the monetary expenses such a project requires. In this case, considerations around cleaning are intermeshed with several other values like work effort, health, sustainability and, of course, nature.

The point which should be stressed here is that cleaning in the natural sciences wavers in its peculiar spheres of value. When it finds its way into interdisciplinary fields, other interests (such as economic ones in the case of the Ganges case study) simply overshadow this particular value regime and project their own values onto debates. This observation of fluctuating value regimes in between disciplines and social contexts demonstrates the field of tension within which cleaning practices take place, are evaluated and effect changes of valuation.

In archaeology, a discussion about the definition, usage and potential of the term and concept of value has predominantly been implied rather than spelled out. The topic was initially hidden behind words like power, prestige or resources. The field of economic archaeology, in particular, (see the fundamental work of Finley, 1973) adapted much of contemporary value theory and plotted it on ancient societies. Questions of personal or moral values are also discussed. Despite the fact that the reconstruction of moral values is highly speculative, archaeologists have been trying to infer social standards and norms from material culture (Bernbeck, 1997:pp.262-263). Further research takes place in the investigation of historical sources, which were handed down to us in time. Since archaeologists have always been tempted to all-too-quickly pick ancient statements that underpin their hypothesis, the awareness of reading the texts from a critical point of view regarding the original authors' intentions is key to applying these sources correctly when it comes to a discussion of value (Meister, 1997; Meister, 1999; Theuerkauf, 1997).

The status of value and evaluating finds is twofold. On the one hand, there may be finds made of bronze, silver, gold, glass, precious stones and so forth. They are considered valuable from academic as well as public perspectives because their base materials are considered rare and valuable in themselves. Here, the material value of the substances that objects are made of is what counts. Archaeology is not the only instance of establishing value in relation to base materials. Its use of material value categories simply reflects historically evolved value systems in the Western world. In the nineteenth century, the first archaeologists were teachers, clergies, army personnel, secretaries of the government or nobles. They projected their contemporary value systems, whether subconsciously or not, on the discovered remains and in turn created ontologies which partly persist until today. The 'Three-Period-System', which, put in basic terms, divides human evolution into the chronological mastering of stone, bronze and iron, exemplifies this (Hansen, 2001). Other factors like a sophisticated production process, limitations due to hazardous trading or means of appropriation may support the definition of material value (Hildebrandt, Neunert & Schneider, 2014:p.239). In such cases, context and social processes around objects make them more valuable than a precious substance as material base. Once a find is highlighted as valuable and has received public and academic attention, it becomes an 'anchor value' and its status is unlikely to change. The bust of the Egyptian queen Nefertiti is such an 'anchor value', which evolves from the sculptural perfection of the bust, the nobility of the depicted and tales from historical sources about the queen's beauty, despite the material of the bust being "just" limestone. The value of this ancient object is preserved, and sometimes even increased, by the ongoing public argument about its rightful ownership between official institutions and states. Whereas the golden mask of the Egyptian King Tutankhamun, which is made of 12 kg pure gold, combines both material and historical value. In contemporary scholarship and popular discourse on the mask, it is therefore framed as a unique example for Egyptian culture.

An example which shows that the material value of an ancient object may be relevant in respect to its potential for re-use is the practice of whitewashing houses in the Mediterranean. People in Greece and Turkey burned marble statues in order to source lime for whitening their walls. These statues were found incidentally or were removed from known ancient sites. In Greece, the historical, aesthetic and symbolic value of the ancient objects was irrelevant for people despite their being descendants of the culture which created them. In modern Turkey, however, some inhabitants reject the history of the Greek city-states on the Asia Minor coast and are therefore particularly motivated to destroy or mistreat possible finds, including sculptural or architectural objects. In both cases, people prioritised the value of the material they wanted to use in everyday practice. Lime ovens, in which marble statues were irretrievably lost, have been common until recent times and they pose a problem to archaeological research. Thankfully, awareness of the importance of these finds has increased enough in contemporary archaeology to encourage local populations to preserve, rather than burn, historic marble objects. These examples show, how differently the (material) value of ancient objects is perceived.

In contrast, "normal" finds, which frequently appear at excavations, may also become interesting for archaeology as objects that should be valued or are found

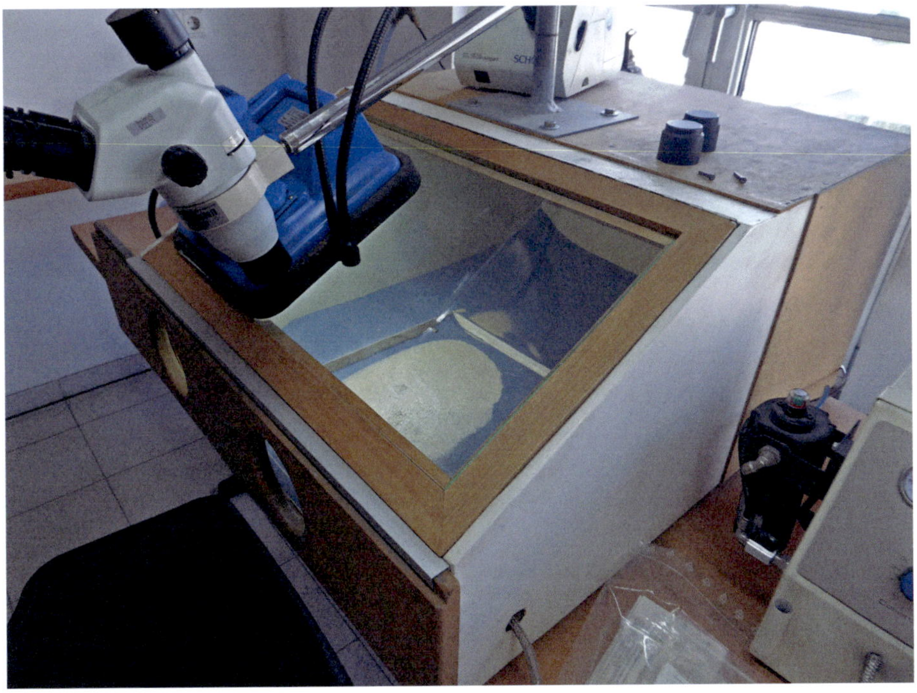

Figure 3. Sandblasting construction from a restauration laboratory (Photo: André Burmann).

Figure 4. View of the box where the objects are processed. At the bottom of the picture one can see the pipe where the sand is emitted by air pressure (Photo: André Burmann).

to be valuable despite being made of a common material. Pottery, for instance, is the most important find category in archaeology due to its omnipresence and its analytical potential for dating, reconstructing trading routes and the economic status of populations, etc. Initially, though, pottery may have been considered of little value compared to an object made out of gold. Things that appear as materially less precious become valuable if they provide researchers with information on the context that is being researched. The informational value, to which objects may become keys, depends on the research question. This, in turn, is influenced by personal, academic, political and societal streams peculiar to each scholar, region and period.

In short, one may say that archaeology, a discipline that is based on working with material traces, prioritises the material value of a find. It does so either through ranking the substance from which it is made on an established scale, either supported by factors that attest to the object's rarity or high craftsmanship or by identifying information within the object's material constitution. Objects only become valuable when the discipline has verified them as valuable through one of these mechanisms. In relation to the previous discussion around the value of cleaning and clean states, archaeology investigates from the point of material culture and substances and may only draw conclusions regarding former practices – which includes cleaning practices – in a secondary step. This assessment, as unspectacular as it may sound in respect to the book's research question, is only half of the truth if you take the process of find restoration into account. Cleaning comes to stand for a mandatory practice, a prelude to investigating excavated material. Interestingly, the value of the base material or the future research question is neglected in this phase because it does not matter whether it is iron, bronze, silver or ceramic, bone or stone. All materials need to be cleaned of earth, rust etc., in order to access the original surface (fig. 3 and 4).

Cleaning finds may generate their value, but it can also be destructive; if, for example, the restorer fails in applying the correct method or the cleaner brushes a shard too intensively, causing the painted surface to vanish undetected.

Discussions of value also hold their proper place in the history of anthropological theory, but, as David Graeber shows, the term has never been developed into one major anthropological theory of value (Graeber, 2001). Rather, different approaches towards proposing such a theory co-exist with various definitions of the term value within anthropological thought. Luis Fernando Angosto-Ferrández (2016) characterises the field, almost poetically, as the 'vast and shape-shifting domain of anthropological takes on theories of value' (p.7). So instead of having a hierarchically structured discourse, there are many discourses – which is not necessarily problematic nor surprising, especially when looking at other discourses around big terms such as, for example, 'myth'. Depending on the nature of the approach that an ethnographic study develops with regards to the life worlds, societies and environments it describes, the kinds of value(s) one may want to focus on will differ. Graeber differentiates, for example, between holistically motivated studies in the structuralist vein, studies that rely on a definition of economic activity as driven by a human desire to maximise, or, among others, studies that intend to consider emotions and experience as part of the package. A Marxist theory of value, which is one of the biggest comprehensive theories in this regard, is used by some anthropologists like Angosto-Ferrández to understand global creations of value. Yet, there are more takes on value in anthropology which may or may not include Marx in their thinking.

In relation to discussing concepts of purity, dirt and cleaning, discussions of value have proven useful for understanding and contextualising studies of waste, recycling and consumption, as Lucy Norris' (2010) or Michael Thompson's (1979) works show. Mary Douglas' work, although bringing considerations around purity and dirt into anthropological discourse, has conceived of value more in the sense of social values. Questions of identifying pure and unclean states are means for building and rebuilding the social order, giving all possible values associated with these processes a distinctly social function. According to David Graeber (2001), these are '"values" in the sociological sense: conceptions of what is ultimately good, proper or desirable in human life' (p.1). Such '"sociological" values, using this term for the sake of simplicity here, may often be found to be rooted in exchange as well as other practices, in things and materials – in short, the stuff that social life thrives on. Materiality, material properties and 'the social relations surrounding these materials' may however also be regarded as decisive indicators for value. As the contributions to Adam Drazin and Susanne Küchler's (2015) edited volume show, materiality may have an impact on the kinds and qualities of values created, be it mediated through an act of making, material properties, regimes of exchange and evaluation or the process of bringing substance into form. While sometimes the social aspect dominates the material aspect, materials may also inform studies which 'represent materials rather than people as the main conveyors of ideas (and) show how materials can produce an exaggerated sense of value around objects'. In comparison to archaeology, anthropology grapples the advantages and disadvantages of a more diverse approach towards value which includes materiality as one aspect without, however, being obliged to prioritise it. As a result, practices that may be identified as cleaning, but also social contexts, as well as material properties, may be factors that determine what cleaning does and what regimes of values it creates, follows or indeed, undermines.

Cleaning and Politics

Given the potential of cleaning practices to maintain, establish, negotiate or revolutionise value, it is not surprising that cleaning practices are also linked to the spheres of politics and power. Examples can be found across different times and in different cultures. A prominent example that demonstrates the connection between power, politics and cleaning practices is the so-called 'Persilschein' (washing powder bill) in Germany after the Second World War (Klee, 1992). The phrase was coined by German soldiers and takes its name from the popular German washing powder brand Persil. Recruits had to bring an empty cardboard box to the military barracks in order to send their civilian clothes home. Often these were Persil washing powder boxes. Hence, the induction order document was dubbed after the detergent: whoever received the call to join the army would have to put their clothes in one of those boxes. After the end of the Second World War, however, the term changed its meaning and became known internationally. During the denazification period, the Allied forces were in the position to issue documents, denazification certificates, which declared their bearers unrelated to the Nazis and their crimes. In order to receive a 'Persilschein' in this new context, one had to provide statements from victims or former enemies that proved one's innocence. Former officers, other members of the government or the public administration could successfully fend off war crime charges if they held a 'Persilschein'. This was a way to secure a place in power and politics in the newly forming democratic state of West Germany. The 'Persilschein' was also necessary to get a business licence.

So the term cleanliness, represented here by the washing powder and by a document, was synonymous with innocence. However, the innocence that was proven may also have been gained by "washing" oneself clean of former misdoings. The 'Persilschein' was connected to the idea of having or regaining a clean slate. In the eyes of the German public, those who received such a document were not at all necessarily innocent, but had instead found a way to buy or threaten their witnesses. In this way, former Nazi officials were able to maintain influence in the new government.

Stefan Solleder (2017) also stresses the political importance of cleaning in the Northern Ireland conflict. Political graffiti was systematically removed. In this volatile period, public space was cleaned from antagonistic slogans which could serve as further political dynamite. A further instance of political cleaning in modern day countries involves monuments in public space which are removed or changed to correspond with the course of recent historical events. This process has happened innumerable times throughout the past in different cultures. It helps to understand the ways in which history and the memory of historical events, persons and developments are created over the course of time. Through acts of cleaning, political players decide upon, and manifest, which memory, which historical figure, is valuable enough to be commemorated and who is to be forgotten.

Conclusion

We have introduced readers to the field of cleaning and value across various disciplines and have found a comprehensive discourse on the role of dirt and its effect on the creation of value. Surprisingly, the practice(s) of cleaning do not receive equal attention. Taking this as a starting point and building on Bruno Latour's critique of modernity, the contributions in this volume test the boundaries of deconstructing strong ontologies. In various ways, they attempt to dismantle the dichotomic distinction of categorising things, people and states as clean or dirty. The (non-comprehensive) inventory of discussions on cleaning and value shows that there has not yet been an attempt to assimilate thoughts around cleaning and its relation to value and, as such, offer a properly interdisciplinary assessment. The contributions in this publication are a step towards this direction, opening up new perspectives on the practical aspect of cleaning and scouting for links between the humanities, the natural sciences and art that exist around this topic. After all, as we hope to demonstrate with the broad disciplinary scope of approaches to this topic, cleaning is something that can be observed and theorised beyond disciplinary boundaries, across time and local contexts. In fact, such an approach is required since cleaning processes point towards a variety of values that inform their assessment and, depending on their context, may lead to very different outcomes. With his critique of modernity as an ideal state that has yet not been, and will not be achieved, Bruno Latour points towards this multi-semantic attribute of the cleaning process as well as the constructedness of the notion of the 'pure' state. Wherever one looks, be it in art, laboratory processes, community organisation or body care, cleanliness remains a precarious condition, at all times threatening to fall back into disorder, dirtiness and the compulsion to clean. As such, it is a Sisyphean task that can never be completed.

Bibliography

Alexander, C. & Reno, J. (eds.) (2012) *Economies of recycling. The global transformation of materials, values and social relations.* London, Zed Books.

Alley, K. D. (2002) *On the Banks of the Ganga. When wastewater meets a sacred river.* Ann Arbor, MI, University of Michigan Press.

Angosto-Ferrández, L. F. (2016) The Value of Everything and the Price of Nothingness. In: Angosto-Ferrández, L. F. & Presterudstuen, G. H. (eds.) *Anthropologies of Value. Cultures of Accumulation Across the Global North and South.* Anthropology, Culture and Society. London, Pluto Press, pp. 1-28.

Arnold, F. (2015) Clean and Unclean Space: Domestic Waste Management at Elephantine. In: Müller, M. (ed.) Household Studies in Complex Societies: (Micro) Archaeological and Textual Approaches. Oriental Institute Seminars (OIS) 10. Chicago. The Oriental Institute, pp. 151-168.

Bernbeck, R. (1997) *Theorien in der Archäologie.* Tuebingen, Francke.

Chakrabarty, D., Trüper, H. and Subrahmanyam, S., eds. (2015a) *Historical teleologies in the modern world.* London, Bloomsbury Academic

Chakrabarty, D., Trüper, H. and Subrahmanyam, S. (2015b) 'Introduction: Teleology and history – nineteenth-century fortunes of an enlightenment project', in Chakrabarty, D., Trüper, H. and Subrahmanyam, S. (eds) *Historical teleologies in the modern world,* London, Bloomsbury Academic

Douglas, M. (1969) *Purity and danger. An analysis of concepts of pollution and taboo.* 2nd ed. London, Routledge and Kegan Paul.

Drazin, A. & Küchler, S. (eds.) (2015) *The Social Life of Materials. Anthropological Accounts of Materials and Society.* London, Bloomsbury.

Duschinsky, R., Schnall, S. & Weiss, D. H. (eds.) (2016) *Purity and danger now. New perspectives.* London, Routledge.

Finley, M. I. (1973) *The Ancient Economy.* Berkeley, University of California Press.

Gell, A. (1992) The Technology of Enchantment and the Enchantment of Technology. In: Coote, J. & Shelton, A. (eds.) *Anthropology, art and aesthetics.* Oxford studies in the anthropology of cultural forms. Oxford, Clarendon Press, pp. 40-66.

Ghanbari, N. & Hahn, M. (eds.) (2013) *Reinigungsarbeit.* Zeitschrift für Kulturwissenschaften 2013 (1). Bielefeld, transcript.

Graeber, D. (2001) *Toward an anthropological theory of value. The false coin of our own dreams.* New York, Palgrave.

Hansen, S. (2001) Von den Anfängen der prähistorischen Archäologie: Christian Jürgensen Thomsen und das Dreiperiodensystem. *Praehistorische Zeitschrift,* 76 (1), 10-23.

Hansman, H. (2015) Scientists find a natural way to clean up oil spills, with a plant-based molecule. *Smithsonian.com.* [Online] 29th July 2015. Available from: https://www.smithsonianmag.com/innovation/scientists-find-natural-way-to-clean-up-oil-spills-with-plant-based-molecule-180955815/ [Accessed 31st October 2018].

Hegetschweiler, J. J. (1833) Versuch zur Beantwortung der von der naturforschenden Schweizerischen Gesellschaft aufgestellten Fragen, die Verwüstungen der Obstbäume durch Insekten betreffend. *Denkschriften der allgemeinen Schweizerischen Gesellschaft für die gesammten Naturwissenschaften, 1 (2),* 65-136.

Hildebrandt, B., Neunert, G. & Schneider, F. (2014) Prestigegüter. In: Samida, S., Eggert, M. K. H. & Hahn, H. P. (eds.) *Handbuch Materielle Kultur*. Stuttgart, Verlag J. B. Metzler, pp. 237-240.

Klee, E. (1992) *Persilscheine und falsche Pässe. Wie die Kirchen den Nazis halfen*. 3rd ed. Frankfurt am Main, Fischer-Taschenbuch-Verlag.

Latour, B. (1994) *We have never been modern*. 3rd ed. Cambridge, Mass, Harvard University Press.

Latour, B. & Woolgar, S. (1986) *Laboratory life. The construction of scientific facts*. 2nd ed. Princeton, N.J., Princeton University Press.

Lossau, N. (2010) Elektrische Reinigung. *Welt-Print*. [Online] 8th September 2010. Available from: https://www.welt.de/welt_print/wissen/article9471481/Elektrische-Reinigung.html [Accessed 1st November 2018].

Markandya, A. & Murty, M. N. (2000) *Cleaning-up the Ganges. A cost-benefit analysis of the Ganga Action Plan*. Oxford, Oxford University Press.

Meister, K. (1997) *Einführung in die Interpretation historischer Quellen. Schwerpunkt: Antike. Band 1: Griechenland*. UTB für Wissenschaft. Uni-Taschenbücher 1923. Paderborn, Schöningh.

Meister, K. (1999) *Einführung in die Interpretation historischer Quellen. Schwerpunkt: Antike. Band 2: Rom*. UTB für Wissenschaft. Uni-Taschenbücher 2056. Paderborn, Schöningh.

Meyer, B. (1999) *Translating the devil. Religion and modernity among the Ewe in Ghana*. Trenton, NJ, Africa World Press.

Meyer, B. (2015) *Sensational movies. Video, vision, and Christianity in Ghana*. Oakland, California, University of California Press.

Norris, L. (2010) *Recycling Indian clothing. Global contexts of reuse and value*. Bloomington, Indiana University Press.

Parker Pearson, M. (2001), The Archaeology of Death and Burial. Texas University Press.

Potters, G. (2015) *Marine Pollution*. [Online] London, Bookboon. Available from: https://bookboon.com/en/marine-pollution-ebook [Accessed 31st October 2018].

Roe, S. (ed.) (2001) *Protein purification applications*. The practical approach series 245. 2nd ed. Oxford, Oxford University Press.

Schmidt, D. (2005) Die Lesbarkeit des Abfalls: Zur Entdeckung materieller Unkultur als Objekt archäologischen Wissens. In: Kienlin, T. L. (ed.) *Die Dinge als Zeichen: Kulturelles Wissen und materielle Kultur. Internationale Fachtagung an der Johann Wolfgang Goethe-Universität Frankfurt am Main 3.-5. April 2003*. Universitätsforschungen zur prähistorischen Archäologie 127. Bonn, Habelt, pp. 239-252.

Solleder, S. (2017) *Die Visualisierung symbolischer Ordnungen im Kontext gewalttätiger Konflikte*. [Online] Berlin, Humboldt-Universität zu Berlin. Available from: https://edoc.hu-berlin.de/handle/18452/19319 [Accessed 3rd December 2018].

Sosna, D. & Brunclíková, L. (2017) *Archaeologies of Waste. Encounters with the unwanted*. Oxford, Oxbow Books.

Stolow, J. (ed.) (2013) *Deus in machina: Religion, technology, and the things in between*. [Online] New York, Fordham University Press. Available from: http://www.jstor.org/stable/10.2307/j.ctt13x095h [Accessed 10th November 2018].

Theuerkauf, G. (1997) *Einführung in die Interpretation historischer Quellen. Schwerpunkt: Mittelalter*. 2nd ed. Paderborn, Schöningh.

Thompson, M. (1979) *Rubbish theory. The creation and destruction of value.* Oxford, Oxford University Press.
van der Loo, M. & de Jonge, E. (2018) *Statistical data cleaning with applications in R.* Hoboken, NJ, Wiley.
Veit, U. (2005/2006) Abfall als historische Quelle. Zeugenschaft in der Archäologie. *parapluie.* [Online] 22. Available from: https://parapluie-de.proxy.ub.uni-frankfurt.de/archiv/zeugenschaft/archaeologie/ [Accessed: unknown (eds.)].
Zimring, C. A. & Rathje, W. L. (2012) *Encyclopedia of Consumption and Waste. The Social Science of Garbage.* Los Angeles, Sage Publications.

Section Two

Disciplinary Reflections

Archaeology and Cleaning: Some Reflections on the Archaeological Process

Ulrich Veit

[1]Within prehistoric archaeology the term or concept of cleaning is present in at least two different contexts. Its first and more important use lies within the realm of field archaeology. 'Cleaning' here represents an activity that forms an integral part of the excavation process, even if the rank of cleaning within a hierarchy of archaeological practices is relatively low. This becomes obvious when looking at the simple fact that cleaning is something normally not performed by the director of an excavation but by workers or excavation assistants. Nevertheless, cleaning has proved vital for archaeological pattern recognition and it requires some experience to produce good results in the cleaning of plana and profiles. In this sense, this aspect of cleaning in archaeology may be relevant from a history of science or laboratory studies perspective.

Apart from this aforementioned context, the reputation of cleaning within archaeology is not the best. On the contrary, cleaning processes in past societies are an obstacle to archaeological work and knowledge production, since cleaning may have destroyed traces that potentially could have allowed conclusions to be drawn about certain aspects of past behaviour and are now no longer accessible to researchers. On the other hand, repeated cleaning procedures in clearly defined archaeological contexts (like houses, settlements etc.) may themselves produce a certain deposition pattern for the archaeological record and therefore these patterns produced through cleaning processes may bear important knowledge about past cultures.

In this respect, one might expect that modern archaeology contributes a lot to a "(pre-) history of cleaning practices". Surprisingly this is actually not the case. I suspect that this results from the fact that archaeologists today still act in the mode of treasure hunters rather than in the mode of trace hunters. In this article I will discuss selected aspects of the overall topic 'archaeology and cleaning' and present some examples. These examples are related to two specifically archaeological aspects that relate to the overall theme of this volume, namely the handling of findings and conclusions at excavations as well as

1 I would like to express may thanks to the organisers of this volume and of the associated workshop not only for the kind invitation, but also for choosing such a demanding topic.

in: Bredenbröker, I., Hanzen, C. & Kotzur, F (eds) 2020: *Cleaning and Value. Interdisciplinary Investigations*, Sidestone Press (Leiden), pp. 41-54.

ethnographies (or better: ethno-archaeologies) of cleaning practices. Since 'cleaning' is neither an established field of research for prehistoric archaeology as a whole nor for myself, the character of my presentation will, to a certain extent, be experimental.

'The term cleaning refers to a conglomerate of practices. These are rooted in social norms, morals and organizational structures. The term includes aspects of materiality, social organization, creation and conservation of value, devaluation as well as destruction'.

This well-chosen formulation from the call for papers may serve as a starting point for my paper. Surprisingly, this statement works equally well if you replace the word cleaning by the word archaeology. Indeed, all the points mentioned here are valid for archaeology as well. But this certainly should not mean that archaeology is the same thing as cleaning: in fact, archaeology is definitely not cleaning per se. The connection between both fields is much more complex. Perhaps one could argue, in a first approach, that the affinity between archaeology and cleaning stems from the fact that both fields are in some sense associated with dirt, waste and refuse.

Indeed, archaeological relics are regularly described in terms related to dirt. A fine example comes from Joachim Reichstein, who compares the archaeologist to a mole. He writes: 'The heraldic animal of archaeologists is the mole. Since [...] Aristotle the opinion exists that this digging small mammal is blind. Certainly not! The snout positioned in the dirt, in his eyes the shine of history.'(Reichstein, 1991:p.38, my translation). Reichstein here refers to the idea that dirt is a constitutive feature of at least field archaeology, but nothing that could prevent an archaeologist from following his idealistic aims.

His message is clear: the archaeologist -in the same way as the mole -does not fear dirt[2]. This clearly is not to say that cleanliness is not as important for moles or archaeologists as it is for other animals or representatives of other professions. Nevertheless, Reichstein's formulation is not unproblematic. Therefore, the metaphoric association of archaeology and dirt has to be questioned -or at least to be qualified.

1.

In my opinion there are at least four qualifications to be mentioned, of which the first one shall be addressed in this introduction.

1. Unquestionably, dust and dirt are concomitants of archaeological excavations. Cleaning practices, therefore, are part of the archaeological process, but they are secondary to the archaeologist's craft. This is proven by the fact that cleaning on excavation sites is normally performed by co-workers or excavation assistants and not by the director of an excavation.
   In any case, excavation is not primarily cleaning but, more accurately, clearing! And in this sense, archaeology since the days of Heinrich Schliemann has often been portrayed as 'the science of the spade' (Veit, 2006a, with further references). Nobody would speak of 'the science of the broom or brush'. Archaeology is known for 'breaking ground' and 'discovering artefacts'- and not for cleaning them[3].

---

2 This fact perhaps can be illustrated by historical photographs from the early days of the discipline. A very spectacular one, that shows only the body and legs of the colleage, with his head hidden in the excavation trench, has been published by K. Hudson (1981, p 64. Fig 16.).

3 Clearly the ‚breaking-ground'-metaphor itself from a more theoretical point of view has ist own problems and is not able to represent the archaeologcial process as a whole. But that is quite another story.

Since it is not possible to cover the topic 'cleaning and cleanliness in archaeological excavation' here in all its facets, I will present only one example that highlights some relevant central aspects. Central to my example is the famous British archaeologist Sir Mortimer Wheeler. Photographs from his numerous excavations all over the world unmistakably demonstrate the high rank of cleaning and cleanliness for his excavation work. Probably no other archaeologist after Wheeler has ever reached such a degree of perfection and it is hard to assess the costs of man-power and organizational, as well as practical, skills required to produce these results.

Wheeler justifies these costs in his famous pocket-book Archaeology from the Earth, which in the mid-twentieth century enjoyed a very wide distribution. Intensive cleaning of archaeological structures by the use of a standard set of tools here is presented as a precondition for archaeological photography and publishing (which in Wheelers time had been largely limited to black and white figures). He writes:

'No amount of mechanical skill is a substitute for the careful preparation of the subject. Clean, sharp angles between the divergent planes of a section, carefully and emphatically cut with the trowel, knife, or edging-tool, are essential if the section is to tell its story with the minimum confusion. Furthermore, a spotlessly clean trench is no mere "eye-wash", if only because it gives the spectator a justifiable trust of orderliness and accuracy of the work. Even the top edges of a trench should be neatly trimmed and the grass cut and swept along them; a stray blade of grass in the foreground of the picture may be overlooked by the eye but may loom embarrassingly in the lens.', Wheeler, 1954:p.200.

Another factor contributing to a perceived cleanliness and order in Wheeler's work is that of leadership and discipline. Both rank high in Wheelers understanding of his profession, which he executed for most of his career in colonial contexts. Wheeler obviously transferred the strict rules of military life to field archaeology. In this sense, the cleanliness on his excavation sites also has to be regarded as an expression of British leadership and discipline. With respect to the high price that the workers involved in these projects had to pay – especially those in the Near and Middle East – one might even think of an obsession for cleanliness which expressed itself in oppressive rule. I will, at this point, refrain from psychological speculations regarding Wheeler himself. I also do not intend to pathologize professional archaeology as a whole in a similar way as it has been done recently in a debate with regards to the inclination of some archaeologists for gathering large collections without any idea of how this mass of material could be of use for archaeological interpretation (Hofmann et al, 2016). Instead it seems vital, to keep in mind the decisive colonial context in which these projects were situated. To a certain degree the structural violence reinforced by colonial governance seems to be reflected in such overly disciplinary behaviour and in the underlying idea of cleanliness.

Let me now proceed in my list of qualifications, which I find necessary to define the relation between archaeology and dirt.

2. Only in very few cases are archaeological objects experienced as dirty by archaeologists, as disgusting and even dangerous in the sense of provoking a fear of physical contact[4]. On the contrary, objects' prolonged presence in the soil -initiating processes of decay, especially of organic materials -may perhaps be regarded as a kind of cleaning. Even archaeological research in medieval latrines is often not as problematic as one might expect from a more removed perspective.

3. It is true that archaeologists mostly deal with refuse, that is to say with things out of use and out of place, which additionally are, to a large degree, fragmented. Nevertheless, these relics are instantly transformed into whole things in the archaeologist's mind -especially into representations of certain object types. The currency of archaeological discourse is not the fragment but items such as strong storage jars, for example, which serve as representatives of overarching artefact types.

4. In the same way, archaeologists that use the term 'settlement refuse' don't normally think of 'refuse' in its strict sense, but rather of the partly unintentional traces of past structured actions. Traces in the soil are taken as indications for a distinctive spatial and social organization of domestic and craft activities.

2.

This kind of archaeological reasoning clearly rests on a semiotic, or 'hunter paradigm' as known from a number of other disciplines like art history or psychoanalysis (Ginzburg, 1988). Such a characterization of the archaeologist as a 'trace hunter' who deals with dirty refuse – and thereby works in a way similar to a criminal investigator – indeed is very popular today, even beyond the discipline.

At the same time, it is far from being generally accepted as the valid representation of archaeological work. Some post-modern archaeologists, including Cornelius Holtorf, have raised objections to this kind of representation (2004:pp.314f.). According to Holtorf, the so-called 'hunter paradigm'is based on false assumptions and cannot therefore be applied to archaeology. But his criticism uses a one-sided interpretation of archaeology as a method for revealing 'historical truth'. This, clearly, is an incorrect interpretation of what the discipline does. I agree with him that an incontestable and unequivocal relationship between causes and effects does not exist. Each trace can potentially be read in very different ways. This is reflected in an old rule of archaeologists: only with intent is it possible to discover what one has been looking for. Put more abstractly one could perhaps conclude: in the end, it is only interpretation which produces the 'trace' and makes it significant.

But this is only one side of the coin. At the same time, 'traces' are a product of past events which had real material impacts. Without such events, archaeologists wouldn't have a chance to construct any 'traces'. So, traces are neither situated exclusively in the past nor in the present but somewhere in between. They are situated between the event and its symbolic interpretation (fig. 1).

Unfortunately, this kind of reasoning has often remained outside of our theoretical reflections related to the past. Indeed, most archaeological theorizing only begins at the point at which a 'fragment' or a 'trace' has been transformed into a man-made object.

---

4 This has been an experience of the Garbage Archaeology of the 1970ies (e.g. Gould/Schiffer 1981).

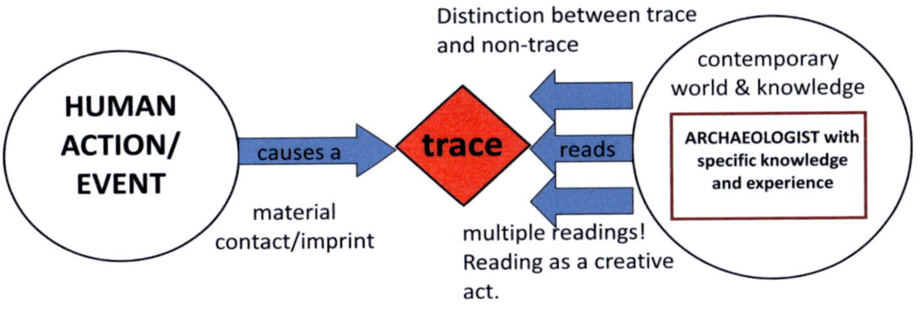

Figure 1. Traces as cultural constructions (Ulrich Veit).

Theories then formulate rules for the way in which these artefacts should be classified and analyzed as members of a special series of artefacts or as elements in special contexts. Round and rectangular buildings, for example, are distinguished as distinctive house types and their spatial and chronological distribution is plotted in order to detect the origin and spread of these types.

The work of the archaeological 'trace hunter' is older and more fundamental as that of the 'typologist'. It involves pattern recognition processes that try to connect single postholes, visible on an excavation plan, to former building types of a specific layout (*e.g.* round or rectangular houses). A similar method is used when trying to analyze the spatial distribution of so-called 'refuse' within dwelling spaces -with the exception that pattern recognition is even more difficult here.

I will come back to this crucial question soon. But first of all, let me sum up our four points concerning the relation between archaeology and dirt.

- Excavations are not about cleaning but about clearing.
- Archaeological objects normally are not experienced as dirty, disgusting and even dangerous.
- The currency of most archaeological discourse is not the relic but whole objects which serve as representatives of certain object types.
- 'Settlement refuse' is not analyzed as 'refuse' but as evidence of structured actions in the past.

3.

These points make it very clear that archaeologists are indeed dealing with dirt, waste and refuse, but they are not the primary focus of archaeological interest. Quite often, they don't even figure as the starting point for formulating questions or conceptualising research. Most archaeological research, up to today, begins its work with a focus on 'treasures' in the common sense of the word. Classic examples could be precious metal hoards and grave inventories with valuable objects, pieces of art or finely crafted objects[5].

5 Hänsel / Hänsel 1997; Wieczorek / Périn 2001. – Another traditional object of archaeological interest in the past have been ‚monuments' in the sense of a special ceremonial architecture (large scale communal buildings for ritual purposes).

Besides these classic treasures, objects of minor material value may also be of particular interest to archaeologists and the public. We could call them 'talking objects' (Daston, 2004) because they evoke a special narrative. These objects may be described by superlatives, for example if they were located at the beginning of an object series (the oldest, largest, heaviest and so on). Some objects stand out as curiosities, as for example a sixteenth century dildo found in a former convent for noble ladies -an object which indeed is very evocative (Landschaftsverband Westfalen-Lippe, 2004:p.61).

For the majority of archaeologists still working in the tradition of nineteenth century positivism, tools and other kinds of everyday objects are the main sources for their interpretations and analyses. Back then, such objects were primarily regarded as fossils of different types with the help of which one could formulate chronological phases. More recently, in extension of this nineteenth century paradigm, archaeological interest has also come to include the traces on objects, documenting special modes of production, use, misuse and even destruction.

The same is true for different kinds of refuse originating from activities related to production, distribution or consumption and even for tiny samples of dust and dirt that have to be analyzed. With the aid of such classifications, we are able to distinguish different ideal methodological types of archaeologies which may be labelled 'monumental', 'typological', 'behavioural' and 'molecular' archaeology. From a history of science perspective, these different types can be associated with different stages in the development of prehistoric archaeology as indicated by the 'invention dates' given here (see box). These dates remain open to debate.

Different Objects of Archaeological Interest – Different Archaeologies

- *Treasures* (e.g. metal hoards & grave inventories with valuable objects)
- *Monuments* (especially large scale ceremonial architecture)
- *'Talking objects'* (curiosities, superlatives)
    - **Monumental Archaeology** (Renaissance - present)
- *Tools and everyday objects*
    - **Typological Archaeology** (1830 - present)
- *Traces* (of actions, production, use, misuse, destruction ...)
- *Refuse* (production / distribution / consumption)
    - **Behavorial Archaeology** (1960 - present)
- *Dust and Dirt* (as well as micro-samples of solid materials)
    - **Moleculary Archaeology / Forensic Science**

To name an example of contesting this chronology, Dietmar Schmidt has argued that the idea of a prehistory developed in the 19th century, primarily as a result of the contemporary scientific discovery of waste (2003; 2004; 2005). For Schmidt, even early prehistory has to be classified as a science of waste. He illustrates his thesis by referring to the discovery and study of the Swiss lake dwellings and the Danish kitchen-midden around 1850. Schmidt refers especially to the famous German pathologist Rudolf Virchow who devoted a considerable amount of time to the study of prehistoric remains (for details see Veit, 2006b).

Indeed, Virchow's theory of the invention of cooking is mainly founded on the archaeological study of ancient kitchen waste. What Schmidt fails to see here is that this new field of research could be described much more adequately as the application of existing geological and palaeontological methods for the analysis of archaeological deposits and less as the discovery of waste through cultural history. Following a logical paradigm, archaeological deposits were analyzed as geological deposits that bear witness to the natural history of man. But this new research paradigm didn't last very long. It was challenged some decades later by the discovery of palaeolithic parietal art, which opened up new perspectives for the archaeological imagination (Veit, 2016).

Prehistoric archaeology as a whole from the early 20th century onwards was dominated by a rather narrow historical culture paradigm with a focus on material evidence of architecture, art and craftwork. Cover illustrations of relevant publications from the time illustrate the evocation of archaeology as treasure hunting (*e.g.* Hoernes, 1892; a cover reprint is to be found in Sklenář, 1983:p.143).

A rediscovery of waste and garbage within archaeology only starts in the second half of the twentieth century, when we see the rise of behavioural archaeology in the United States. With the adoption of the idea of taphonomy a new interest in geology and palaeontology becomes evident.

4.

The changing attitudes of archaeology to dirt and refuse may be perhaps best illustrated by turning towards one of the key sites of modern archaeology and its excavators. The site of Troy (western Turkey) is inextricably connected to the names of Heinrich Schliemann, Wilhelm Dörpfeld, Carl W. Blegen and Manfred Korfmann. While Schliemann became famous for his – probably to a large degree constructed or even faked – treasures (such as Primos) and the historic narratives he presented to the public, Dörpfeld was especially successful in studying architectural remains. While their work remained important during the 20th century, new perspectives developed. To name one example, Carl Blegen in particular attempted to introduce a new perspective to dirt by systematically analyzing and interpreting the debris-fillings of the Troy ruins (Blegen et al, 1950; Blegen et al, 1951).

Carl Blegen interpreted the thick infills of EBA period III buildings – labelled alternatively as 'debris', 'rubbish', 'garbage' or, more neutrally, 'floor deposits'-as a result of careless housekeeping. He writes:'The accumulation of debris within the houses shows the people of Troy III to have been careless housekeepers: in the course of time the floors became littered with refuse shells, animal bones, sherds, and decomposed matter; when the garbage became unbearable, a new floor of fresh clean earth or clay was laid down, only in its turn to be covered with rubbish.' He later added: 'When a floor became offensively encumbered with rubbish of animal bones, shells, potsherds, and other litter, it was covered by a new floor of fresh earth or clay which in its turn suffered the same fate [...]. At intervals, interrupting the succession of floors, thicker masses of clay, perhaps fallen from the roof, perhaps brought from outside, indicated a more serious readjustment of levels.'(citations taken from Blum, 2002:pp.105, 108f.)

In the light of new theoretical insights and excavations, however, this kind of ad hoc interpretation has to be questioned. Referring to common sense, Blegen clearly underestimated the complexity of archaeological formation processes as demonstrated by modern behavioural archaeology (*e.g.* Schiffer, 1987; Sommer, 1991). According to behavioural archaeology, different types of refuse as, for example, secondary refuse or abandonment refuse are distinguished. Contrary to the so-called 'Pompei Premise', the archaeological record could not be regarded as a snap shot of a specific moment in the past (cf. Binford, 1981; Schiffer, 1985).

With these fresh ideas in mind, Stephan Blum, a member of the Korfmann-excavation team, re-analysed refuse management and archaeological fillings of Early Bronze Age Troy (2002). He postulates that the buildings excavated by Blegen were indeed cleaned from everyday refuse on a regular basis during their time of use. According to Blum, the archaeological fillings -as visible to the archaeologist -were mainly composed of refuse and debris that began to form after the buildings had been abandoned:

> *'(...) during the abandonment of buildings, almost all things left within the spaces are useless things. Objects that can be regarded as typical of the living culture are commonly not part of archaeological contexts or severely underrepresented in them. Cultured objects found on the floors of buildings are predominantly objects which only reached the location where they were excavated after the active use phase of the respective buildings, be it as abandonment, de facto, secondary or tertiary refuse. They therefore may not correspond with the activities that were carried out in these locations under normal circumstances.'*, Blum, 2002:pp.137.

This insight has consequences for archaeological interpretation far beyond the site of Troy. It shows that ancient behaviour-including cleaning practices -can hardly be understood by means of conventional settlement archaeology. In most studies, the archaeological record does not reflect daily routines but rather occasional events, especially those connected to the abandonment of whole settlements or single archaeological features -as for example storage pits, which normally contain all kinds of materials, sometimes even human skeletons (Müller-Scheeßel, 2013). Without a careful study of post-depositional transformations of the archaeological record we are unable to draw sound historical conclusions (for a further debate of these question see also Sosna & Brundlíková, 2017, part 3).

5.

From an archaeological perspective, one is confronted with a somewhat ontological problem when attempting to understand ancient mentalities and ideologies concerning dirt and waste. In contradiction to the convictions of some behavioural archaeologists, dirt and waste are not universal concepts. They are cultural categories which, in their current form, only emerged during the nineteenth century. Seen from a historical perspective, refuse is a by-product of modern infrastructures like refuse collection or waste water circulation (van Laak, 2017).

In the contemporary Western world, floor and road surfaces are a major factor in preventing the resurfacing of dirt and waste. Unlike in the past, dirt and waste don't just disappear unnoticed into the soil by becoming a 'natural' part of it. In the same way, faeces in urban contexts are no longer recycled locally or collected for reuse (van Osten, 2016), but are dealt with by an elaborate and expensive waste water infrastructure. This shows that our understanding of dirt and waste is not directly applicable to pre-modern, and especially to prehistoric conditions.

Recycling is another key word of modern industrial waste management, that is today regularly also applied to pre-modern archaeological contexts. But in the case

Figure 2. Garden installation, Eastern Friesland, 1985 (Photo: Ulrich Veit).

of these early cultures the underlying idea must have beenquite different from the modern idea of waste management. We are facing here a special form of object-related behavior, that has to be distinquished from our modern mentality to replace repairable or still usable objects quickly by new ones – because this is convenient and not expensive. Durables in the prehistoric past – as well as in other pre-modern contexts – were used as long as possible or they were disposed of for ritual purposes (Hänsel & Hänsel, 1997). When things lost their function, the materials from which they were made may have been used in a variety of secondary contexts.

Similar practices of re-use may be found in the recent past and, in certain cases, even today (Ludwig-Uhland-Institut für empirische Kulturwissenschaft der Universität Tübingen & Württembergisches Landesmuseum Stuttgart/Volkskundliche Sammlung, 1983). This point is illustrated by the work of an unknown Frisian 'artist' that combines a characteristic element of modern waste water infrastructure with a particular sense for practical value and aesthetics (fig. 2).

For those familiar with modern art, especially Marcel Duchamp's ready-mades (*La Fontaine*), this piece of popular art may evoke some earlier debates about the question 'what is art' and what is valued as art? This concept of value is vital for understanding the role of cleaning in an archaeological context, since cleaning (or recycling) practices never serve practical reasons alone. They are always also an expression of ideals of cultural order, of purity and pollution. Questions that deal with dirt and cleaning always possess a social and a moral dimension (see also Reno, 2017).

Taking this point into account, dirt and cleaning may become the topic of a social archaeology of prehistoric times. For the moment, however, it seems as if truly convincing case studies in this field are still amiss (nevertheless, see: Sommer, 1991; Sommer, 1998; Fansa & Wolfram, 2003 and recently Sosna & Brundlíková, 2017). Instead, a large number of studies are situated in the field of historical or contemporary archaeology, using a broad spectrum of data.

Figure 3. Graffito at Zürich/Switzerland, 1982 (Photo: Ulrich Veit).

This paper does not claim to give a detailed review of prehistoric or historic case studies. I will mention, therefore, one example of a promising research field for an archaeology of the contemporary. Urban graffiti is a specific element of modern visual culture and a subject of the public discourse around dirt and cleaning. Graffiti is therefore an interesting class of evidence for a social archaeology of modern society. Because their production normally remains invisible to the public -it mainly happens at night -the attempt to interpret graffiti resembles, to some degree, the work

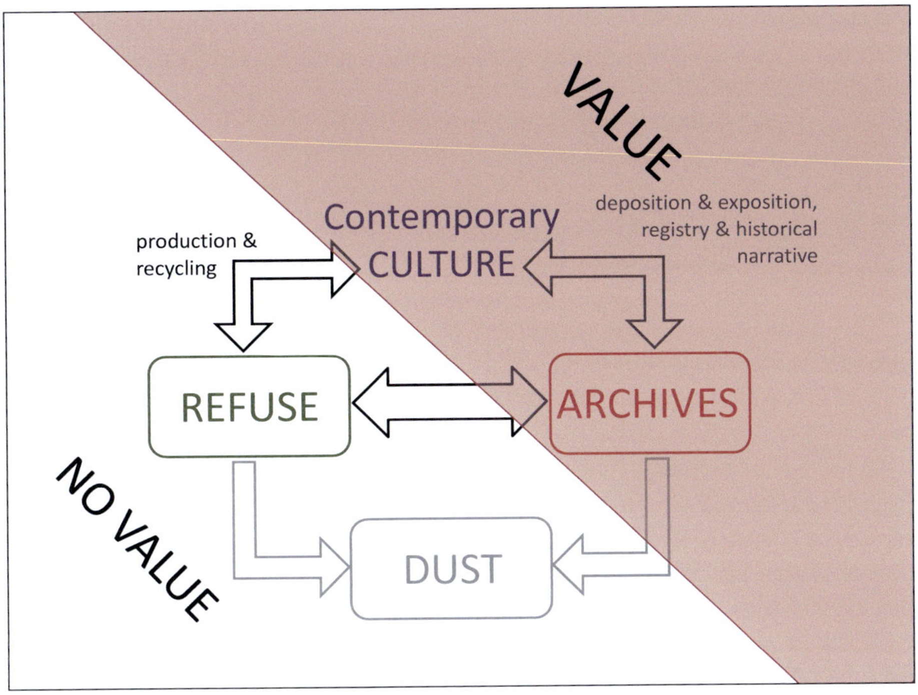

Figure 4. Archaeology and the production of value (Ulrich Veit).

of archaeologists. Some standard archaeological methods, like typography, stylistic analysis and even stratigraphy, may be useful in this context. Possible key elements for such an analysis would certainly be the specific locations (accessible surfaces with high visibility), the diverse techniques of application and, of course, the textual and pictorial elements used for the graffiti -sometimes including ironic comments with regards to their urban context and contemporary society (fig. 3). At the same time, their existence is always at risk, since these kinds of public statements may be seen as pollution and destruction.

Conclusion

I hope that I have been able to raise at least some of the questions that are of relevance for the overall topic of this volume with regard to archaeology and especially to prehistoric archaeology. It should have become clear that a straight equation of archaeology and cleaning is problematic. Instead, archaeology could perhaps be more accurately described as a practice -among others -involved in the transformation of the state of material things (fig. 4).

In the case of archaeology it is mainly the transformation from ‚invisible' to ‚visible', from ‚dark' to ‚bright', from ‚forgotten' to ‚effective', and finally from ‚refuse' to ‚valuables' that were stored and exposed. We could also speak of a transformation of ‚refuse' to ‚symbols' in the sense of a (re-)activation. One might perhaps be inclined to add a further pair of words to this list: dirty and clean. But as I have tried to demonstrate, at least with regards to archaeology, this binary metaphor does not work very well.

Postscript

The author would like to thank the anonymous reviewers for their critical comments and valuable suggestions on an earlier draft of this paper as well as the anonymous person(s), who have helped to make accessible my English also to native speakers.

Bibliography

Binford, L. R. (1981) Behavioral Archaeology and the 'Pompeii Premise'. *Journal of Anthropological Research*, 37 (3), 195-208.

Blegen, C. W., Caskey, J. L., Rawson, M. & Sperling, J. (1950) *Troy. General Introduction. The First and Second Settlements. Volume I.* Princeton, Princeton University Press.

Blegen, C. W., Caskey, J. L. & Rawson, M. (1951) *Troy. The Third, Fourth, and Fitfth Settlements. Volume II.* Princeton, Princeton University Press.

Blum, S. W. E. (2002) Vom Hausfleiß der Troianer. In: Aslan, R., Blum, S., Kastl, G., Schweizer, F. & Thumm, D. (eds.) *Mauerschau.Festschrift für Manfred Korfmann. Band 1.* Remshalden-Grunbach, Bernhard Albert Greiner, pp. 105-151.

Daston, L. (2004) Speechless. In: Daston, L. (ed.) *Things that Talk. Object Lessons from Art and Science.* New York, Zone books, pp. 9-24.

Ebeling, K. & Altekamp, S. (eds.) (2004) *Die Aktualität des Archäologischen. in Wissenschaft, Medien und Künsten.* Frankfurt a. M., Fischer Taschenbuch.

Fansa, M. & Wolfram, S. (eds.) (2003) *Müll. Facetten von der Steinzeit bis zum Gelben Sack. Begleitschrift zur Sonderausstellung. Müll -Facetten von der Steinzeit bis zum Gelben Sack vom 06. September bis 30. November 2003 in Oldenburg anschließend in Hanau.* Schriftenreihe des Landesmuseum für Natur und Mensch 27. Mainz, Philipp von Zabern.

Ludwig-Uhland-Institut für empirische Kulturwissenschaft der Universität Tübingen & Württembergisches Landesmuseum Stuttgart/Volkskundliche Sammlung (eds.) (1983) *Flick-Werk. Reparieren und Umnutzen in der Alltagskultur. Begleitheft zur Ausstellung im Württembergischen Landesmuseum Stuttgart vom 15. Oktober bis 15. Dezember 1983.* Stuttgart, Württembergisches Landesmuseum Stuttgart.

Ginzburg, C. (1988) Spurensicherung. Der Jäger entziffert die Fährte, Sherlock Holmes nimmt die Lupe, Freud liest Morelli – die Wissenschaft auf der Suche nach sich selbst. In: Ginzburg, C. (ed.) *Spurensicherungen. Über verborgene Geschichte, Kunst und soziales Gedächtnis.* 2nd ed. Munich, Deutscher Taschenbuchverlag, pp. 78-125.

Gould, R. A. & Schiffer, M. B. (eds.) (1981) *Modern Material Culture. The Archaeology of Us. Studies in Archaeology.* New York, Academic Press.

Hänsel, A. & Hänsel, B. (Bearb.) (1997) *Gaben an die Götter. Schätze der Bronzezeit Europas. Ausstellung der Freien Universität Berlin in Verbindung mit dem Museum für Vor- und Frühgeschichte, Staatliche Museen zu Berlin – Preußischer Kulturbesitz. Bestandskataloge Bd. 4.* Berlin, Staatliche Museen zu Berlin -Preußischer Kulturbesitz, Freie Universität, Seminar für Ur- und Frühgeschichte.

Hoernes, M. (1892) *Die Urgeschichte des Menschen nach dem heutigen Stande der Wissenschaft. Vienna,* A. Hartleben's Verlag.

Hofmann, K. P., Meier, T., Mölders, D. & Schreiber, S. (eds.) (2016) *Massendinghaltung in der Archäologie. Der Material Turn und die Ur- und Frühgeschichte.* Leiden, Sidestone Press.

Holtorf, C. (2004) Archäologie als Spurensicherung. In: Ebeling, K. & Altekamp, S. (eds.) *Die Aktualität des Archäologischen. in Wissenschaft, Medien und Künsten.* Frankfurt a. M., Fischer Taschenbuch, pp. 306-324.

Hudson, K. (1981) *A Social History of Archaeology*. The British Experience. London, MacMillan.

Landschaftsverband Westfalen-Lippe (ed.) (2004) *Das Museum. wma. westfälisches museum für archäologie. landesmuseum*. Münster, Landschaftsverband Westfalen-Lippe.

Müller-Scheeßel, N. (ed.) (2013) *'Irreguläre' Bestattungen in der Urgeschichte: Norm, Ritual, Strafe ...? Akten der Internationalen Tagung in Frankfurt a. M. vom 3. bis 5. Februar 2012*. Kolloquien zur Vor- und Frühgeschichte 19. Bonn, Habelt.

Reichstein, J. (1991) Das archäologische Denkmal als archäologische Quelle. In: Horn, H. G., Kier, H., Kunow, J. & Trier, B. (eds.) *Archäologie und Recht. Was ist ein Bodendenkmal?*. Schriften zur Bodendenkmalpflege in Nordrhein-Westfalen 2. Mainz, Philipp von Zabern, pp. 31-38.

Reno, J. (2017) Wastes and Values. In: Sosna, D. & Brunclíková, L. (eds.) *Archaeologies of Waste. Encounters with the unwanted*. Oxford, Oxbow, pp. 17-22.

Schiffer, M. B. (1985) Is There a 'Pompeii Premise'in Archaeology? *Journal of Anthropological Research*, 41 (1), 18-41.

Schiffer, M. B. (1987) *Formation Processes of the Archaeological Record*. Albuquerque, New Mexico, University of New Mexico Press.

Schmidt, D. (2003) „Kommt Zeit, kommt Unrat". Abfallforschung und die Entdeckung der Vorgeschichte im 19. Jahrhundert. In: Fansa, M. & Wolfram, S. (eds.) *Müll. Facetten von der Steinzeit bis zum Gelben Sack. Begleitschrift zur Sonderausstellung. Müll -Facetten von der Steinzeit bis zum Gelben Sack vom 06. September bis 30. November 2003 in Oldenburg anschließend in Hanau*. Schriftenreihe des Landesmuseum für Natur und Mensch 27. Mainz, Philipp von Zabern, pp. 187-195.

Schmidt, D. (2004) Abfall und Vorgeschichte. Entdeckungen der Prähistorie im 19. Jahrhundert. In: Ebeling, K. & Altekamp, S. (eds.) *Die Aktualität des Archäologischen. in Wissenschaft, Medien und Künsten*. Frankfurt a. M., Fischer Taschenbuch, pp. 263-282.

Schmidt, D. (2005) Die Lesbarkeit des Abfalls. Zur Entdeckung materieller Unkultur als Objekt archäologischen Wissens. In: Kienlin, T. L. (ed.) *Die Dinge als Zeichen. Kulturelles Wissen und materielle Kultur. Internationale Fachtagung an der Johann Wolfgang Goethe-Universität. Frankfurt am Main 3.-5. April 2003*. Universitätsforschungen zur prähistorischen Archäologie 127. Bonn, Habelt, pp. 239-252.

Sklenář, K. (1983) *Archaeology in Central Europe. The First 500 Years*. Leicester, Leicester Universiy Press.

Sommer, U. (1991) Zur Entstehung archäologischer Fundvergesellschaftungen. Versuch einer archäologischen Taphonomie. In: Mattheußer, E. & Sommer, U. (aus dem Seminar für Vor- und Frühgeschichte der Universität Frankfurt/M.) *Studien zur Siedlungsarchäologie I*. Universitätsforschungen zur Prähistorischen Archäologie 6. Bonn, Habelt, pp. 51-193.

Sommer, U. (1998) Kulturelle Einstellungen zu Schmutz und Abfall und ihre Auswirkungen auf die archäologische Interpretation. In: Schmidt, M. (ed.) *Geschichte heißt: So ist's gewesen! abgesehen von dem wie's war ... Geburtstagsgrüße für Günter Smolla*. Archäologische Berichte 11. Bonn, Habelt, pp. 41-54.

Sosna, D. & Brunclíková, L. (eds.) (2017) *Archaeologies of Waste. Encounters with the unwanted*. Oxford, Oxbow.

van Laak, D. (2017) *Archäologie des Alltags. Köln und seine Infrastruktur*. Cologne, Greven.

van Osten, R. (2016) Nightman's muck, gong farmer's treasure: local differences in the clearing-out of cesspits in the Low Countries, 1600-1900. In: Sosna, D. & Brunclíková, L. (eds.) *Archaeologies of Waste. Encounters with the unwanted.* Oxford, Oxbow, pp. 41-56.

Veit, U. (2006a) Mehr als eine ‚Wissenschaft des Spatens'. Troia und die Geburt der modernen Archäologie. In: Korfmann, M. O. (ed.) *Troia. Archäologie eines Siedlungshügels und seiner Landschaft.* Mainz, Philipp von Zabern, pp. 123-130.

Veit, U. (2006b) Gründerjahre: Die mitteleuropäische Ur- und Frühgeschichtsforschung um 1900. Foundation Years: Pre- and Protohistoric Research in Central Europe around 1900. In: Callmer, J., Meyer, M., Struwe, R. & Theune, C. (eds.) *Die Anfänge der ur- und frühgeschichtlichen Archäologie als akademisches Fach (1890-1930) im europäischen Vergleich. Internationale Tagung an der Humboldt-Universität zu Berlin vom 13.-16. März 2003. The beginnings of academic pre- and protohistoric archaeology (1890-1930) in a European perspective. International Conference at the Humboldt University of Berlin March 2003, 13-16.* Berliner Archäologische Forschungen 2. Rahden, Westfalen, Marie Leidorf, pp. 43-61.

Veit, U. (2016) Leo Frobenius und die ‚Kunst der Vorzeit'. Neue Forschungen zur Entdeckungs- und Wirkungsgeschichte prähistorischer Felsbilder. *Ethnographisch-Archäologische Zeitschrift*, 55 (1/2, 2014), 210-218.

Wheeler, R. E. M. (1954) *Archaeology from the Earth.* Oxford, Clarendon Press.

Wieczorek, A. & Périn, P. (eds.) (2001) *Das Gold der Barbarenfürsten. Schätze aus Prunkgräbern des 5. Jahrhunderts n. Chr. zwischen Kaukasus und Gallien.* Publikationen des Reiss-Museum 3. Stuttgart, Konrad Theiss.

How is Dirt possible? On the Philosophy of Dirt, Cleanliness and Refuse

Olli Lagerspetz

To ask *how is dirt possible* is to ask: what are the conditions of human thought and life that make it meaningful to use the concept of dirt at all? When attempting to answer this question, related concepts like those of refuse, pollution, soiling and cleaning will also have to be addressed. What is involved in applying these and similar descriptors to material objects, and what does the fact of their application imply about our ordinary relations with our physical surroundings? As Hans Peter Hahn points out[1], the role of dirt-related concepts has to do with what it means to *assign value* to objects, or perhaps, as I would prefer to put it, with what it means to *recognise* the values that objects already have.

The question 'how is dirt possible?' is of course an allusion to Immanuel Kant who framed some of his central enquiries in this form. He asked, among other things, how synthetic *a priori* truths were possible, how mathematics and pure (*i.e.* theoretical) natural science were possible, and how the categorical imperative was possible. These questions assume that a certain phenomenon or practice, such as pure natural science, clearly exists. There is, however, something about our other philosophical or intellectual commitments that implies that it somehow ought not to be possible. Kant, for instance, argued that a workable conception of laws of nature, and thus of pure natural science, could not be upheld if philosophy remained committed to the then received idea of the relation between *a priori* and *a posteriori*. Given that pure natural science – which so to speak should not exist – in fact is a meaningful undertaking, how should we revise our intellectual commitments?

It seems to me that dirt is in a similar kind of predicament. Given some of the intellectual commitments typical of academic culture at present, it may seem that dirt does not fit in. Existing debates on dirt, soiling and impurity are, to a great extent, attempts to come to grips with a perceived incoherence between the phenomenon of dirt and our commitments. The most important of these commitments is a certain view on the relation between physical reality and culturally determined ideas about reality. Natural science is expected to be the ultimate arbiter of the real and the unreal. Whatever falls outside

1 In his opening remarks for the conference for which this essay was written.

in: Bredenbröker, I., Hanzen, C. & Kotzur, F (eds) 2020: *Cleaning and Value. Interdisciplinary Investigations*, Sidestone Press (Leiden), pp. 55-66.

its reach belongs to the domain of mere subjectivity. Clearly, our descriptions of things as dirty, soiled, clean and the like imply *a dimension of meaning or value* which seems incompatible with science as we today understand it. The crucial idea was expressed concisely by Justus von Liebig more than a hundred years ago: *'Für die Chemie gibt es keinen Dreck'* (for chemistry, no turd exists).

Scientific Realism, Culturalism and Aristotle: a first Approximation

The structure of my argument is as follows. In theoretical analyses of soiling we find two main approaches, labelled here for short as scientific realism and culturalism. Less prominent today, there is the possibility of a third approach, connected with Aristotle. These perspectives represent different ways of making intellectual sense of the phenomenon of dirt or, as the case may be, of claiming that the phenomenon itself is illusory. The bulk of this essay is devoted to developing the last of these three approaches. The two others must, however, be outlined in order to get an idea of the current state of the debate.

Scientific realism as a philosophical position comes in many varieties, but now it is sufficient simply to indicate its general outlook. Its main theses are, firstly, that reality is independent of the mind and, secondly, that the entities and qualities that really populate the world are the ones that would be picked up by the predicate terms of a 'mature' or fully developed (natural) science (see Agazzi, 2017:pp.209f.). Historically, scientific realism emerged when natural science was enthroned as the preferred framework for rational inquiry of reality. The aim of science, according to scientific realism, is to present descriptions of reality independent from any subjective point of view. To quote Thomas Nagel, it is to present 'a view from nowhere' (Nagel 1986). For realism, objective reality exists, but adherents of scientific realism argue that everyday experience alone cannot give us an undistorted view of reality (Agazzi, 2017:p.210). Some key aspects of reality are *inaccessible* to the untrained mind and liable to be misconstrued by it. Therefore, in scientific realism, theoretical natural science is presented as the best approximation to a true and completely subject-less account of reality. In such a perspective, dirt tends to disintegrate, sucked up into the general category of physical substance.

'Culturalism' is, for the most part, not a term used by the thinkers who are subsumed in that category. Nevertheless, it may be a good term to describe a tendency that is widely accepted in anthropological approaches to pollution.[2] Dirt is understood as a subjective, symbolic and culturally conditioned product of the mind. In a sense, the approach which is here labelled culturalism can be regarded as the opposite of scientific realism, because it focuses precisely on the experiences of human subjects. On the other hand, culturalism and scientific realism spring from a common root and they can be made mutually compatible. If scientific realism is accepted as the best approach to reality as such, we are left with a 'remainder' of human experience that seems to require treatment of some other kind. Material reality as such is handed over to science while its subjective aspects are represented as the business of psychology and cultural anthropology. In culturalism, dirt becomes, in the words of Mary Douglas (1970), a result of 'the differentiating activity of the mind' (p.190). It is viewed as something that the mind imposes on an essentially neutral reality. Dirt belongs to a layer of symbolic meanings attached to objects.

2 The term 'culturalism' was suggested by a reader of the present volume.

If the question is, 'how is dirt possible?', the answer from the perspective of culturalism and scientific realism is, then, that strictly speaking dirt *is not* possible; not as a *real* feature of the world. Dirt falls outside of science and, while *ideas* of dirt exist in culture, they have no objective validity.

There is, finally, a third perspective that promises the chance to side-step the dichotomy between scientific realism and culturalism. This is an approach inspired by the Aristotelian distinction between substance (consisting of matter and form) and accident. In this essay, I hope to make the case for such an approach. The argument is twofold. On the one hand, one can generate an argument based on the shortcomings of the two other approaches. On the other hand, I argue that the alternative approach captures everyday experience where the two others do not.

When, during the early Modern Age, Galilean and Newtonian physics replaced the earlier Aristotelian conception, one central change was that differences between *kinds* of material things were no longer considered. There was no difference of principle, in Galileo's thinking, between living and lifeless objects, nor between natural objects and artefacts. For Aristotle, contrastingly, different theoretical concepts were appropriate for accounting for different types of objects. Ultimately, Galileo's physics proved to be more conducive to scientific development, but Aristotle's view has, in the present context, the advantage of being more closely modelled on the immediate human experience of living in a material environment.

Revisionary and descriptive Metaphysics of Properties

The argument in this essay is thus based on the assumption that dirt *is* possible. A philosophy that claims the opposite would self-disqualify, simply because it would not be an analysis of the concept of dirt but a denial of its applicability. This methodological commitment is based on the idea of philosophical analysis as a descriptive enterprise. If philosophy is descriptive, its preferred aim is to make sense of concepts, not to explain them away, hereby getting rid of them. The analysis of a concept requires us to attend to its applications and hence, to start by identifying *meaningful* instances of its use. The crucial question for the descriptive analysis is not, do the words 'dirty' and 'clean' correspond with real qualities in the world, but rather: *given that* we use these words in order to relate to the material environment, what are we able to learn about the implicit assumptions made with regards to this environment which guide our thinking and acting?

My chosen approach involves a choice between what Peter Strawson (1959:p.9) identified as *revisionary* and *descriptive* metaphysics. Descriptive metaphysics aims to analyze the conditions and presuppositions of knowledge and understanding as they appear in various contexts of enquiry. Revisionary metaphysics, instead, is not content with conceptual analysis but aims to uncover the true ontological structures of reality as such. It is treated as an open question whether our most quotidian categories for describing ordinary objects, human action, etc. truly identify anything which exists. In the anglophone philosophical discourse of the last three or four decades, the role of metaphysics is almost universally perceived as revisionary both by its defenders and its detractors (D'Oro, 2012).

Both realism and culturalism are species of revisionary metaphysics because they attempt to determine whether dirt really exists, and in that case, what it objectively *is*. They are open to the possibility that 'dirt' stands (or is meant to stand) for something that is not really there. It might also be possible to interpret Aristotle's metaphysics as revisionary. I suggest, however, that it is better understood as an attempt to *articulate* the thinking that is already characteristic of our awareness of the material environment.

'Aristotelianism' is not the only possible articulation of our life world in a way that side-steps the 'subjective vs. objective' dichotomy.[3] Such a striving is present, for instance, in the constructivism put forward by Bruno Latour (2013). Evaluating constructivism falls outside the present essay, but the generally descriptive stance seems to be a feature that it shares with Aristotle.

The Idea of Dirt as Projection

The idea of dirt as a projection is the starting point of culturalism as described above. Material objects in themselves are neither clean nor dirty, but there are human beings who project their emotions and normative expectations on objects.

This way of conceptualising our relation with material reality is evident, for instance, in Sigmund Freud's *Totem and Taboo (1946, 1948).* In that book, Freud attempts to account for our perception of 'the uncanny' (*das Unheimliche*). Our perception of various objects and chains of events as uncanny, he says, is due to what he calls *the outward projection of inner perceptions*. He describes such a projection as a process whereby

> *"inner perceptions of ideational and emotional processes are projected outwardly, like sense perceptions, and are used to shape the outer world, whereas they ought to remain in the inner world."*, Freud, 1946:p.85f.

When we ascribe the property of uncanniness to an object or to a chain of events, we are in the grip of a kind of magical thinking. We incorrectly expect our psychological states somehow directly to modify the environment. It is a subjective colouring of an originally colourless world.

More recently, Julia Kristeva (1982:p.60) in her book *The Powers of Horror*, quotes this passage from Freud with approval. She applies it to the human perception of dirt and pollution. At the centre of Kristeva's discussion of dirt lies the concept of 'abjection'. Dirt is defined by our reactions of rejection and disgust, the ultimate aim of which, she believes, is to safeguard the integrity of the subject as a separate individual and a separate body.

Let me just note in passing that there are certain risks about placing too much emphasis on the role of disgust – or of any emotion – in our perception of dirt. 'Dirty' and 'disgusting' certainly do not mean the same thing. It is safe to say that, for most people, disgust is not a universal or even dominant reaction to dirty objects.

It would be odd to characterize Kristeva as an adherent of scientific realism even though that description might be applicable to Freud. Nevertheless, it is interesting to note that the idea of *an essentially neutral world*, a real world not captured in everyday perception, is something both Freud and Kristeva *share* with scientific realists. Kristeva takes up this very same idea of dirt as an expression of an emotional state, a state which we project upon a world that is in itself neutral.

3 For example, Bruno Latour and Helmuth Plessner explicitly strive to overcome similar dichotomies. This point was raised by an anonymous reader of this chapter. See Latour, 2013; Plessner, 1975. Latour can be read as presenting the descriptive point that the idea of objective reality is a thinking tool. As such, the functions of the concept of reality are to be articulated in exactly the same way as those of other critical concepts. It is not an ontological master concept.

Shoes on the Table and 'in themselves'

To put it briefly, the idea of dirt as a human emotional and cultural projection is this: if dirt is not physical, then it must be a projection. The most famous example of this approach comes from Mary Douglas (1970), originally from 1966. In a passage that has become a *locus classicus* in the research, she contrasts 'shoes in themselves' with 'shoes on a dining table':

> *"Shoes are not dirty in themselves, but it is dirty to place them on the dining-table; food is not dirty in itself, but it is dirty to leave cooking utensils in the bedroom, or food bespattered on clothing; similarly, bathroom equipment in the drawing-room; clothing lying on chairs; out-door things in-doors; upstairs things downstairs; under-clothing where over-clothing should be; and so on. In short, our pollution behaviour is the reaction which condemns any object or idea likely to confuse or contradict cherished classifications."*, Douglas, 1970:p.48.

Douglas draws the conclusion that dirt, like beauty, is 'in the eye of the beholder'. She defines dirt as 'matter out of place'; and what *is* out of place in a given case is determined by a *symbolic world order* characteristic of the culture in question. Anything that appears to challenge order or deviates from it, will risk being perceived as impure and dangerous. Throughout her book, Douglas pursues an argument that assimilates the ordinary concept of dirt to the general concept of anomaly in the context of a symbolic order. Douglas has been (and still is) enormously influential upon research related to pollution taboos and cleanliness in culture. In that research, descriptions like 'dirty' and 'soiled' are typically associated with social categories like 'the forbidden' rather than with material qualities like 'wear and tear', 'wet', 'rusty' or 'damaged'.

A lot could be said about the example of 'shoes on a table'. First of all: when Douglas states that 'it is dirty' to place shoes on the table, we are easily convinced because, on hearing the example, we naturally think of *dirty* shoes on a table, not of clean shoes straight out of the box. *Clean* shoes on a table may certainly (sometimes) count as 'matter out of place' and hence as messy or untidy, but we would not typically see them as dirty – and certainly not treat them as *dirt* even if they satisfied the definition of 'matter out of place'. This example, and the other examples in the quoted passage, are (or at least can be) cases of disorder; no doubt. But Douglas has not shown that dirt is a species of disorder. You can tidy up a room without cleaning it, and you can clean a room without tidying it.

On the other hand, something of central importance is certainly brought out by Douglas in the quoted passage. When we think of dirt we must also think of human involvement. In the words of Edwyn Bevan, 'in an uninhabited world moist clay would be no dirtier than hard rock; it is the possibility of clay adhering to a foot which makes it mire' (quoted in Ashenburg, 2007:p.279). And we must think of a culture, in this case one where shoes are used to protect one's feet when walking. Our understanding of what counts as soiling on a shoe, and our understanding of how shoes can be ruined, are connected with our understanding of the characteristic situations in which shoes are used.

However, the contrast which Douglas makes between shoes 'in themselves' and shoes 'on the dining table' may be misleading. Shoes in themselves, she says, are not dirty. But to this one could reply: if there indeed is such a thing as a 'shoe in itself', then it already implies the human practice of walking. An object outside of those practices is not a shoe. We will then not be speaking of a shoe but of an undefined material object of rubber and leather. And a culture where shoes are used for walking inevitably involves practices of caring for one's shoes, protecting them against damage and soiling, as well as tending to their repair and cleaning. All of which implies a richer and more context-bound conception of material things than is allowed by any clear-cut dichotomy between the subjective and the objective. The rest of this essay is meant to outline what this richer conception of material things might amount to.

Substance and Accident

Objects around us bear the stamp of human needs and values. Consider the fact that almost everything in our everyday milieu, as it now exists, is the result of conscious human efforts to reshape the environment. That is true, more or less, for any element of a normal living space. This fact also means that we can easily imagine different kinds of disturbances, ways in which artefacts around us might deviate from their proper conditions. Any adequate description of a man-made environment is likely to involve a perspective where the contrast between the ideal case and deviations naturally comes in. Soiling is one such deviation.

For a more in-depth analysis of the concepts of dirt and soiling, it will be helpful to hark back to the Aristotelian distinction between substance and accident. For Aristotle, a substance is any self-sustained thing such as a stone, a cat, a teacup or a human being. (Thus, the Cartesian definition of material substance as simple extension is foreign to the Aristotelian conception of substance.) The substance is essentially a combination of matter and form. In this scheme, an accident is a quality that does not affect the fundamental identity or essence of the substance. For instance, the colour of a teacup may be seen as an accident: if the cup is bleached out, the substance still survives. Colour, in its turn, cannot exist alone without some underlying substance. The colour of a human being is also an accident, whereas her essentially human qualities, such as rationality and two-legged body structure, make up the human form which, when instantiated in matter, constitutes her as the specific substance she is. We should, however, note that the distinction between substance and accident is open for interpretation. How we specify the distinction in a given situation is dependent on what we perceive to be essential for the identity of the substance. In some cases, colour is an important element of the substance. A painted picture is destroyed if the colour is removed.

Aristotle believed that his 'matter versus form' and 'substance versus accident' distinctions were applicable to any physical substance. In that sense, it is unsurprising that we can also apply them to dirt and soiling.[4] The more interesting question is: what aspects of our lived experience are highlighted or obscured if we do so?

The original distinction of substance and accident highlights the difference between qualities that essentially belong to an object and those which are somehow added onto it. The identity, essence or substance of the object may be summed up in the description of its 'normal state', which here means its rightful, normatively correct state. Accidents like dirt, damage, wear and tear are secondary. Dirt does not change the essence of the underlying substance, which can be made visible again, as when you remove a stain. Damage is a different kind of modification, because it implies that the substance itself is affected. According to Aristotle, it is possible for a substance to be damaged or mutilated

---

4 As pointed out by a reader of the present volume.

if the configuration of its parts is essential to its form. For a human being, having a leg cut off results in mutilation, whereas substances like water or wax can increase or diminish without an essential change in substance (Ross, 1960, 1023b-1024a). The concepts of damage and dirt are similar, however, such as that they imply a contrast between the present state of an object and its underlying form or essence.

Philosopher Thomas Leddy (2012) makes use of precisely this distinction between substance and accident in his paper on what he calls 'everyday surface aesthetic qualities' (1995:p.259 and *passim*). He describes 'dirty' as 'a *surface* quality'. By this he does not just mean that dirt collects on the surfaces of objects. For instance, a liquid may be thoroughly dirty. Similarly, in the case of greasy hair, you cannot typically point to dirt on a delimited part of its surface; it is the hair's general condition that counts. Nevertheless, these judgements involve the general act of distinguishing between a given substance as such and whatever is foreign to it. Here we are implying a relation between two unequal factors: a master object – a shoe, for instance – and an additive – clay, for example. For Leddy, 'dirty' is a surface quality insofar as it can be kept *analytically distinct* from the fundamental 'underlying form or substance' of the master object. To clean an object or tidy up a space is to reveal a form which has been clouded by unessential additions.

Thus the background assumption in our judgements about soiling must be that the master object is in principle *possible* to clean, that it in some sense *needs to* be cleaned and is *worth* cleaning (Leddy, 1995:p.260). Perhaps this is the reason why bits of toilet paper are not typically described as dirty but simply as 'used'. We do not think there is an underlying substance worth cleaning; cleaning would in any case hardly be practically possible. Used toilet paper is called dirty mainly when there is a danger that it may soil *other* objects. The normative position outlined here implies a judgement concerning the relative values of the (valuable) master object and the (worthless) additive. On the other hand, it does not always require a fixed set of priorities. Consider another example: food falling on a carpet. If food falls down it may ruin the carpet, but at other times we say, conversely, that food is ruined when it falls on the carpet.

These descriptions imply a hierarchical relation between the master object and the additive, between substance and accident. The master object is treated as valuable or interesting in its own right while the additive is reduced to its role as a disturbing element. In a sense, dirt in this scheme is not a substance at all, but a kind of disturbance that affects an existing substance. If you isolate a sample of dirt and analyse it on its own, it becomes something else: a chemical substance in its own right. In this sense, being 'dirty' or 'soiled' is like being 'wet'. 'Wetness' occurs when an object makes contact with water. Water certainly exists as a substance, but it becomes 'wetness' only in contact with a master object. One does not say water is wet except in connection with the idea of something or someone coming in contact with water.

Considerations of this kind distinguish dirt from certain other unwanted elements such as trash, refuse, rubbish, garbage and faeces. Unlike dirt, these elements are substances in their own right. They are discarded, not because they touch another object and ruin it, but because of what these elements themselves are. A 'trashy' object *is* trash or it is *like* trash, but a dirty object is not itself dirt. On the contrary, the implication is that the dirty object needs cleaning precisely because it is something *different* from dirt. This is, incidentally, a distinction not honoured in a number of influential theoretical accounts of the concepts of dirt and impurity (see Bataille, 1970; Douglas, 1970; Kristeva, 1982; Nussbaum, 1999). For instance, Julia Kristeva (1982) writes of the dead human body: '[t]he corpse, seen without God and outside of science, is the utmost of abjection' (p.4). Her description contrasts starkly with traditional practices of washing the dead. The body is washed, indicating precisely that a human corpse is valuable. It *may* be cleaned and it is *worth* cleaning.

Caring for Objects

The upshot of the Aristotelian argument, as presented so far, is this: the key to our conception of dirt consists in our ability to recognise the 'everyday' *identities* of given objects. This recognition includes awareness of differences between what belongs to the object and what should count as an alien, accidental or disturbing addition. This is, in a sense, a normative conception of dirt because it involves an understanding of what it is for the object to be in the *right* way and in the *wrong* way. Words like 'dirt' and 'soiling' belong among a range of words that denote a departure from the desirable ideal state of an object, a milieu or a living being. We may therefore say: to be familiar with an object and to know what kind of object it is involves, among other things, understanding what would count as an unacceptable kind of soiling of it. Sometimes dirt has practical significance, as for example when grime and build-up clog a washing machine drain. Very often, however, our concern with dirt has to do with purely aesthetic considerations.

In some cases, an object is called dirty not out of concern for the object itself, but because of the need to protect other objects. This is typically the case with human hands. When you ask me, "Are your hands clean?", what is usually of interest is not the state of my hands as such, but that I should adequately handle an object you care about. The right answer to your question will be dependent of the character of the object in question. My answer is an expression of my idea of what it means to handle that specific object with care.[5]

It seems to me that disagreements about what constitutes soiling in a given case may be quite often traced to differences in opinion about the nature of the master object. The interesting case of the conservation of books is described by Anna Magdalena Lindskog Midtgaard (2006) who works at the Rare Books section of the Copenhagen Royal Library. Major libraries today have custom made vacuum cleaners for books, and there are also techniques for washing and ironing book pages. Some librarians find it important to remove stains and dust from old volumes, thinking of the *new* volume as the ideal. Others would take a more conservational approach. Grains of pollen and sand may be seen as belonging to the volume's history. They sometimes contain useful information about the volume's place of origin and the hands through which it has passed up to its present location. This is in many ways similar to a typical situation in archaeology, as highlighted by Ulrich Veit (this volume). At the excavation, the archaeologist faces questions about what to clear away and what to keep as part of the archaeological findings. The variety of existing attitudes among librarians not only reflects differences in taste, but also ideas about the identity of the item itself. A stain on a book may either be seen as a blemish or as patina: either as something external to the volume or as a natural feature of it. Technically speaking, patina is impossible to distinguish from wear and dirt, but the

5 I was once asked what is the dirtiest object in the world. This is, of course, an impossible question, but since a child was asking, I felt obliged to come up with an answer. I said, "the human hand", because it is the object that most frequently needs washing.

description of it as 'patina' implies that it would be an act of vandalism to remove it. The old manuscript volume should convey the message, 'I am 500 years old'; but it must not necessarily cry out, 'I was *new* 500 years ago'. On the other hand, all librarians would probably agree that it is desirable to remove additions like Sellotape, which actually causes much damage to the paper underneath.

The challenges of maintaining a book collection highlight one more aspect of our understanding of the concept of soiling: the idea of a responsible attitude to one's environment – something Susan Strasser (1999:p.21) has called our 'stewardship of objects'. Not only *we* have demands on our environment but, conversely, the objects around us have demands on us. To understand what kind of object an old manuscript volume is involves understanding what kind of proper care and handling it requires. In this way, the world unfolds itself to us as a set of possibilities and requirements. The volume requires being handled with caution; my shoes require cleaning; and these requirements exist independently of us as individuals (Sartre 1962:p.39). In sum, dirt is possible because a sense of *responsibility* is integral to our understanding of what an everyday object is. We recognise the difference between objects in their clean or ideal state and in their disturbed state. And we see that it is someone's responsibility to restore or protect the ideal state from which soiling is a deviation.

Conclusion

To ask 'How is dirt possible?' is to ask, 'What are the conditions under which this kind of awareness of the environment is meaningful?'. Our everyday concepts of dirt and soiling belong to our ongoing interaction with a humanly shaped environment. Through this interaction, material elements reveal themselves not only in the form of neutral physical entities but as things with distinct identities. The identities of everyday objects are made manifest through the various ways in which things can go wrong with them.

In moral philosophy, living beings are sometimes described as entities that have a '*welfare*' or a '*well-being*' (Crisp, 2017). Living beings have needs that call for attention; living beings can be treated ill or well. The fact that living beings have a welfare is perhaps seen most clearly when they suffer. We can, for instance, immediately tell when a potted plant has been neglected. The plant must be watered, not because someone wants it that way but because, as a living thing, the plant has a welfare. One way to sum up the argument in the present essay is to say that so-called 'lifeless' objects also may have a welfare. Things can go well or badly for them, and they need attention from us. It is thus plausible to say that human thinking, in addition to specifying the categories of living beings and 'mere' objects (as in physics), also counts on a third category: that of objects with a purpose built into their identities. This creates the framework for a language and a life in which objects can be described as damaged and mended, disheveled and tidy, dirty and clean.

One will note that this analysis can be generalized beyond just the question of soiling. The general intellectual consensus in the global West has been that the world 'in itself' is mute and empty of meaning – its magic is gone, it is 'disenchanted', as Max Weber famously put it. However, the disenchantment thesis does not correspond to our experience as human individuals. We are born into a world where objects always already have purposes, waiting for us independently of any ideas that we might personally have about them. This is the aspect of reality that neither scientific realism nor culturalism have captured adequately.

In the everyday experience of an individual, ordinary objects are purposeful as a matter of course. Our everyday awareness of the material environment contains an Aristotelian element: that of a distinction between substance and accident. Without contradicting the previous point, it also contains elements of Platonic thinking. Our experience of the world is shaped by ideas of perfection and of falling short of that perfection.[6] Our experience is internally structured by a notion of value or of the Good – not in opposition to facts but itself a condition of the meaningful perception of facts.[7]

Bibliography

Agazzi, E. (ed.) (2017) *Varieties of Scientific Realism. Objectivity and Truth in Science.* Cham, Switzerland, Springer.

Ashenburg, K. (2007) *The Dirt on Clean. An Unsanitized History*. New York, North Point Press.

Bataille, G. (1970) La valeur d'usage de D.A.F. de Sade. In: Bataille, G. *Oeuvres complètes II. Écrits posthumes 1922-1940.* Paris, Éditions Gallimard, pp. 54-72.

Bevan, E. (1921) *Hellenism and Christianity.* London, George Allen Unwin Ltd.

Crisp, R. (2017) Well-Being. In: Zalta, E. N. (ed.) *Stanford Encyclopedia of Philosophy*. Fall 2017 Edition. [Online] Stanford, California, Metaphysics Research Lab, Stanford University. Available from: https://plato.stanford.edu/archives/fall2017/entries/well-being/ [Access: 6 July 2020].

D'Oro, G. (2012) Reasons and Causes: The Philosophical Battle and the Meta-philosophical War. *Australasian Journal of Philosophy,* 90 (2), 207-221.

Douglas, M. (1970) *Purity and Danger. An Analysis of Concepts of Pollution and Taboo.* Harmondsworth, Penguin Books.

Freud, S. (1946) *Totem and Taboo. Resemblances between the psychic lives of savages and neurotics.* Authorized translation with an introduction by A. A. Brill. New York, Random House.

Freud, S. (1948) *Totem und Tabu. Einige Übereinstimmungen im Seelenleben der Wilden und der Neurotiker.* In: Freud, A., Bibring, E., Hoffer, W., Kris, E. & Isakower, O. (eds.) *Sigm. Freud. Gesammelte Werke, Chronologisch geordnet. Neunter Band. Totem und Tabu.* 2nd ed. London, Imago Publishing.

Hämäläinen, N. (2014) What is a Wittgensteinian Neo-Platonist? – Iris Murdoch, Metaphysics and Metaphor. *Philosophical Papers*, 43 (2), 191-225.

Kristeva, J. (1982) *Powers of Horror. An Essay on Abjection.* New York, Columbia University Press.

Latour, B. (2013) *An Inquiry into Modes of Existence. An Anthropology of the Moderns.* Cambridge, Massachusetts, Harvard University Press.

Leddy, T. (1995) Everyday Surface Aesthetic Qualities: "Neat," "Messy," "Clean," "Dirty". *The Journal of Aesthetics and Art Criticism,* 53 (3), 259-268.

Leddy, T. (2012) *The Extraordinary in the Ordinary. The Aesthetics of Everyday Life.* Peterborough, Ontario, Broadview Press.

---

6 This is independent of whether we want to postulate Platonic Forms as independently existing – the kind of 'two world theory' that is traditionally often associated with Platonism.

7 Thus, according to Nora Hämäläinen (2014), our understanding of facts includes or points towards a hierarchy of value; a 'dynamic principle of our lived, everyday experience' (p.217), which 'is not of our own making, and which places demands on us' (p.215).

Midtgaard, A. M. L. (2006) *The Dust of History and the Politics of Preservation*. Paper for the Nordic Summer University Winter Symposium, Circle 4: Information, Technology, Aesthetics, 3rd -5th March 2006, Helsinki.

Nagel, T. (1986) *The View from Nowhere*. New York, Oxford University Press.

Nussbaum, M. C. (1999) "Secret sewers of Vice". Disgust, Bodies, and the Law. In: Bandes, S. A. (ed.) *The Passions of Law*. New York, New York University Press, pp. 19-62.

Plessner, H. (1975) *Die Stufen des Organischen und der Mensch. Einleitung in die philosophische Anthropologie*. Sammlung Göschen 2002. 3rd ed. Berlin, Walter de Gruyter.

Ross, W. D. (ed.) (1960) *The works of Aristotle. Volume III. Metaphysica.* 2nd ed. Oxford, Clarendon Press.

Sartre, J.-P. (1962) *Sketch for a Theory of the Emotions*. Translated by Philip Mairet. With a preface by Mary Warnock. London, Methuen & Co Ltd.

Strasser, S. (1999) *Waste and Want. A Social History of Trash*. New York, Metropolitan Books.

Strawson, P. F. (1959) *Individuals. An Essay in Descriptive Metaphysics*. London, Routledge.

Section Three

Contexts and Places

The Last Bath: Cleaning Practices and the Production of 'Good Death' in an Ewe Town

Isabel Bredenbröker

How to Clean Up After Death?

The connection between a dead body and cleaning can seem deceptively obvious. At the end of a person's life, the corpse is a decomposing material remainder which must be dealt with and "cleaned up". Death is frequently associated with pollution of physical and spiritual kind (Hertz, 2004; Huntington & Metcalf, 1979). The corporeal body is subject to different practices of cleaning which both factor in, and exceed, its materiality: washing, preservation, presentation or concealment and spiritual cleansing. Although it might be tempting to deduce a natural connection between death and cleaning, ethnography proves this to be a premature conclusion. Whilst all human beings die, death is not universally experienced or symbolically framed (Bloch & Parry, 1982; Hertz, 2004; Huntington & Metcalf, 1979). Rather, from the perspective of the living, death gives expression to a diverse range of practices: it is culturally and socially produced. This diversity of practices and beliefs shatters the preconception of death as the denominator of a shared human condition. Ways of what I, following Robert Hertz, will call "making death" within the social sphere are culturally and locally specific but may to an extend be comparable or serve similar ends, albeit in different ways. Cleaning practices form one element of how people deal with death and they may differ in their execution, their publicness or hiddenness and their participants (Jindra & Noret, 2013). My paper examines the entanglements of cleaning practices -in relation to death as well as in other parts of life – and the organization of political, spiritual and social authority in a Ghanaian community. As the ethnographic material that I gained from fieldwork over 8 months between 2016 and 2018 shows, social organization, death and cleaning practices are mutually connected and form a nexus which highlights the mechanism of distributing power in the community.

In Peki, a small town in the Ghanaian Volta Region that is populated largely by people of Ewe ethnicity, death and cleaning practices coincide and play particular roles as publicly enacted events. Mourning and funerary arrangements include, among other events, the

in: Bredenbröker, I., Hanzen, C. & Kotzur, F (eds) 2020: *Cleaning and Value. Interdisciplinary Investigations*, Sidestone Press (Leiden), pp. 69-88.

Figure 1. Obituary banners hanging in town, alongside presidential election campaign posters and flags as well as advertisement for a church. (Photo: Isabel Bredenbröker).

semi-public washing of the body, huge publicly held funeral celebrations and lying-in-states, public visual commemoration of the dead on 'obituary posters', the joint effort of digging graves and specific pre-conditions for receiving a burial permit. These events and practices are connected to a structure of community labor, payments and expenses in relation to death, moral evaluations of the deceased and spiritually informed practices such as traditional-religious rituals or Christian funeral services.

Community labor involves, among other tasks, the cleaning of the town, for example picking up rubbish or emptying clogged waste-water drains in public space. On top of that, there are also hygiene controls imposed by the Ghanaian state that check facilities in family homes such as toilet and waste water management, waste deposition and drinking water storage. These controls are executed by a local group of representatives and workers in the service of the Environmental Health Unit, under the auspices of the District Assembly of South Dayi District to which Peki belongs. Districts in Ghana, with their assemblies, departments and units form a locally represented, second-level administrative structure. The people who execute the controls in town are a unit manned with town folk which is embedded in the national governance structure. Environmental Health Officers are working for the national Ministry of Health. Such hygiene checks cut across the boundary of the private household, combining state regulations with local actors. Through hygiene controls as well as through communal labor, both state-imposed as well as communally organized, it is public knowledge who cleans, who lives cleanly and who does not. Equally, people in town know who is active for and in the community, who attends community labor or holds a voluntary position and who does not. By partaking in communal labor, community members can achieve a reputation as active members of the community. The fact that people know who is active in the upkeep of the community translates into how deaths are processed in the community and how different states of cleanliness are achieved – on social and spiritual levels and in relation to rules imposed by the Ghanaian state. The moral status of a deceased person is produced in a communal evaluation process with a multiplicity of participants who may have contradicting views. How a person is evaluated to have performed as a community member influences the conditions under which one can be washed, laid in state and buried – in short, the conditions for properly reacting to their death. These evaluations are, among other things, related to questions of cleanliness and what is morally acceptable.

Instances in which cleaning and other practices of dealing with death intersect in the Peki community stand at the center of this paper. I take Mary Douglas' considerations on dirt as a point of departure. Douglas defines the elimination of dirt, in other words cleaning, as a 'creative effort to organize the environment' (Douglas, 1966). However, Douglas' lack of attention to practices, in the sense of a praxeological approach, and her treating the concept of order as monolithic prevents her from adequately thinking 'the multiplicity of relations of power and forms of categorization in society' (Duschinsky, 2016). In this case, practices that I focus on are, among others, cleaning practices in the mortuary context as well as within the everyday lives of community members. Building on the aforementioned discourse around dirt and cleanliness, I understand cleaning as a set of practices that are expressed in direct interaction with the material environment, allowing people to negotiate divergent ideas of what is clean and what is dirty by means of a common material ground. As I will show via ethnographic description, practices at the intersection of cleaning and death exist within a larger context of community organization and politics. They are sites of power struggles and serve to negotiate the intentions and status claims of different parties, revealing such a 'multiplicity of relations to power' (Duschinsky, 2016). Questions of spirituality, the social rules of the local community and state rules equally play into these processes of negotiating power(s).

The Role of Death in Town

The town of Peki consists of several sub-communities with individual township names, united under the umbrella term of Peki. In pre-colonial times, Peki was organized as a state under the reign of a traditional chief and its territory extended beyond what is today the town of Peki incorporated in the state of Ghana. The traditional political system of chieftaincy is specific of Ghana and has remained in place as a governing structure alongside the governance of the nation state. Today, each Peki sub-community has a chief and there is a paramount chief as well as elders and chiefs of clans and families. For the entire community, death plays a major role in the proceedings of public life. Large scale obituary banners and smaller posters cover the surfaces of houses and are prominently propped up on long sticks at the town entrance, like a welcome committee. Fridays mark the beginning of multiple funerals in town. The second day of my fieldwork already catapults me into an evening procession along the town's main street in which mourners dance frenetically, play music and fill up the street, pushing people back into the entrances of houses and blocking cars which carry the bodies of the deceased.

Figure 2. Environmental Health Unit workers during their shift in town as they walk through a neighbourhood inspecting houses. (Photo: Isabel Bredenbröker).

Practices in the funerary cycle

Over the course of my fieldwork, I took part in multiple funerals and other activities related to death, mourning and funerary events. Funeral weekends are organized on a bi-weekly schedule, marking the first or second weekend of a month and the third or fourth. The calendar with the official funeral weekends is published by a representative of the Peki Traditional Area, the *adontehene*, by appointment of the paramount chief or *deiga*.[1] With the exception of two weeks after New Year's Day, funerals happen all year round. On Friday mornings, undertakers arrive at the morgue with a delegation of family members to perform a washing of the body. The morgue is situated within the grounds of the Peki Government hospital. Both institutions, hospital and morgue, are formally supervised, staffed and financially administered by the Ghanaian state; the state, however, is here represented by the local hospital administration, just as in the case of the people forming the local Environmental Health Unit.

1 Burials on non-funerary weekends are allowed when an extra fee is paid to a community account under the supervision of the local chief and a chosen group of elders.

After washing, the body of a deceased remains in the morgue until the afternoon. Bodies may only leave the hospital grounds and enter populated parts of town after sunset due to a ruling by the representatives of the Peki Traditional Area: chiefs and other traditional council members, who all ensure compliance with traditional religious rules. As stated in a set of rules pinned to the morgue wall, issued by the hospital, these bodies are not to be picked up from the morgue later than 6 P.M. This combination of state-imposed and traditional rules makes timing essential in the process. Cars might have to wait on the way to town or drive very slowly if they are early. Late pick-ups may have to happen in a hurry. The transportation of bodies into town can be accompanied by groups of mourners of various size, depending on a person's popularity, age and cause of death. In town, the body is received by a delegation of the deceased's maternal and paternal family who sit opposite each other at an agreed meeting point and exchange gifts of schnapps, pouring generously to the ancestors while doing so. Multiple checks occur to verify the identity of the body. Then a lying-in-state at a family member's house follows for which the body is dressed by the undertaker and put on display for mourners who can interact with it.[2] People can speak to the deceased, touch them, put gifts in the coffin, sing to them, tell them not to come too close, cry and pray. Saturdays are funeral days, usually marked by a service in the house or at a church, after which the body is taken to a cemetery and buried. Sunday mornings are used for family gatherings to solve possible disagreements that may have occurred over the course of the weekend. This schedule is a rough map for different funerals taking place at the same time over a weekend. The culmination of a synchronized sequence of events means that funeral weekends are intense for the whole community and engage the wider public as well as guests coming from Accra or elsewhere in different roles: as guests, as workers, as mourners as well as in religious, economic or spiritual activities.

Spirituality and the Temporality of Death

Mary Douglas connects dirt to liminality and danger (Douglas, 1966; van Gennep, 1977). Due to death's liminality as a passage between worlds, some anthropological work frames it as a spiritual pollutant which needs to be cleaned up in order to reinstate the balance of everyday life (Kim, 2012; Lamont, 2013). In Peki, the interconnectedness of several religious practices – different varieties of Christian churches, traditional religious practice as well as a Muslim minority – results in an interesting pluralistic interpretation of death from a spiritual point of view.

A majority of the town's population are Christians, alongside a small Muslim community and practitioners of local traditional religion. A multitude of churches – Presbyterian, Pentecostal and other charismatic churches – distinctly shape the face and soundscape of the town. These strands of Christianity are a blend of Christian and local traditional religious beliefs as Birgit Meyer has shown in her comprehensive ethnographic work on the symbolism of the devil in Peki (Meyer, 1999). Both coexist. Following Christian beliefs does not mean that traditional beliefs, such as belief in spirits of ancestors and nature deities, are negated. Rather, they are acknowledged but may not necessarily be

---

2 On rarer occasions, for example in cases of particularly bad causes of death such as suicide or murder, bodies may not be laid in state at all or placed in a public square. Privately organised funerals may on occasion also hold their laying-in-state at a public place or building.

Figure 3. Gifts for the spirits of deceased who died a bad death at an *agbadome* site. (Photo: André Burmann).

engaged with actively. There are, however, exceptional situations. After a young man in town died in a car accident, his Christian family, facing the second accidental death of a family member, decided that a traditional ceremony in which his soul was collected from the place of accident was necessary to prevent further misfortunes. The soul, in the form of soil from the accident site, was deposited in a white cloth and carried on an elder's back like a baby. Everyone who took part in the picking up received a strip of the cloth that had to be worn around the wrist on the way back. The soil and all the cloth were then buried in the grave of the deceased. During a second part of the ceremony that was held after sunset, luggage such as food, pots, clothing and cosmetics were placed in an *agbadome* site, a place for 'luggage' objects for the spirits of the accidentally deceased. These rituals serve as protective acts for the community. The collection of the soul helps to prevent further accidents from happening, either at the same site or to relatives. The deposition of luggage is supposed to satisfy all the needs of the spirit for a life in between the worlds so it will not come to bother the living. At the same time, as I have witnessed, the collective march to pick the soul up and further celebrations at the family house also serve as a sort of outlet for communal aggression, excessive drinking and grieving. The death of a young person and violent deaths are reasons to be angry and to display this anger as a shared emotion.

Regarding the spirits of the deceased, the coexistence and crossover of beliefs creates the paradoxical situation that these spirits could exist in several states of afterlife at the same time. A soul could be believed to be in heaven or hell and, at the same time, in a place where ancestors dwell, or as a wandering spirit in concrete locations such as *agbadome*. The belief that deceased relatives become ancestors when some time has passed is, in conversation with different community members, mostly referred to in a quite vague way. Ancestors appear to be generally considered as a presence that dissolves individual existence. The most commonplace practice in relation to ancestors is the pouring of alcoholic drinks as a sign of sharing and respect towards them. Depending on different factors such as a community member's political agenda, their educational background, religious affiliation or age, views on the relevance of ancestors and their whereabouts may differ. For example, a person expecting to become a catechist in a local church may publicly refrain from pouring libations or showing themselves at events associated with traditional beliefs, while still believing that both ancestors and spirits do in fact exist. The same multiplicity of views (and ways of behaving in the face of them) exists in relation to the presence of wandering spirits, as in the case of those who have died in a bad way, such as death by accident, murder or suicide. It is likely, however, that people do not completely negate the existence of any of these forms of the afterlife, but rather prefer to actively endorse one and not the others. The blend of traditional and Christian religion is, however, omnipresent as can be seen in the biography of a local traditional priest who changed her calling from church member to traditional practitioner.

The distinction between the world of the dead and the world of the living is not clearly defined, which means that they can mutually influence each other (Locke, 2016). Death, in different religious beliefs alike, becomes a non-finitude, a state of existence, which changes the place and manifestation of a person but does not mean that they cease to exist. Temporality also changes in death, as it is understood here. Time becomes a permanent present, transgressing boundaries between spaces and worlds. The frozen body and a preference for solid, synthetic materials which do not decompose nicely underline the local imagination of death as a state of present tense. In all parts of the mortuary cycle, there is a preference for choosing materials and objects made out of synthetic materials over locally-sourced, organic equivalents. Grave wreaths, to name just one example, are made of plastic foil, synthetic gift ribbon and a cardboard base, wrapped in cellophane.

The idea of permanence is confronted by the material reality of death as a transformative process, including decomposition of the body and a physical disappearance of the animated person. Just as death is a transformative and liminal state, so can matter find itself in multiple states. Matter does not know death, only change. New Materialist approaches frame matter as something in permanent vital transformation with a multitude of properties that may be active or inactive (Bennett, 2010; Coole & Frost, 2010; Drazin, 2015). The difference between dead matter and human subjects becomes ineffective. Instead, potentially vital matter also makes the death of human subjects less 'deadly'. The local practices around death in Peki aim at redirecting forms of change towards a more stable eternal present tense, sometimes making use of change as a quality in things, at other times opting for materials which only change to a limited degree. Material things and substances related to death need to balance a fine line between controlled or permitted change and permanence. The body may not rot uncontrolled – a vitality of matter which in this context does not translate well into the idea of a permanent present tense. Instead, cleaning practices and the preference of non-organic materials both work against unwanted change – or dirt – and come to produce and represent a kind of 'eternal life' in death. Organic matter and the change which it undergoes has the potential to represent that same temporality, as in the example of the soil-soul or represented by the use of cassava plant sticks instead of headstones on graves. In these instances, organic materials are of crucial importance after all and do play an important role as memory objects and carriers of symbolic value. But preferably, people in Peki opt for synthetics and materials which, at first glance, do not seem to 'die' like natural flowers.

Figure 4. Synthetic grave wreaths on finished 'rich' graves and developing grave plots at a cemetery in Peki. (Photo: Isabel Bredenbröker).

In the face of a permanent presence of the dead in multiple worlds and places, cleaning up spiritually calls for different practices, which address this distributed present tense of death. While Christian services take care of the soul and its passage to heaven, funerary practices also require other kinds of cleaning processes which relate to the ancestral existence of a deceased or their presence as spirit. The interrelation between Christian and traditional beliefs means that there is no hard distinction between these practices. Instead, they connect the spiritual realm with more worldly aspects of social representation. A place like a cemetery can address several spiritual realms and transport messages of social valuation at the same time through different material aspects and uses. Cemeteries and the materiality of graves communicate ideas of distinction as well as of longevity through a choice of synthetic and durable materials – some of which are imported and harder to get or more expensive. An expensive grave that claims a certain status for both relatives and

Figure 5. Obituary banner out of tarpaulin with digital colour print standing in as headstone and grave cover on a 'poor' grave, held in place by cassava plants. (Photo: Isabel Bredenbröker).

the deceased may be constructed from bricks, cement and a solid cover with an engraved granite headstone. A 'poor' grave may just be decorated with a cassava stick or a cactus into which the name of the deceased has been scratched by hand.

Cemeteries also have an economic value as places which create income for the community and individuals. Payments for the use of graves contribute the main share of money which is saved for the community in the communal bank account. Economic activities such as the making and selling of obituaries, coffins, wreaths and other things associated with funerals provide sources of income and business opportunities for people. *Agbadome* sites and accident sites are regarded as transitory places where the spirits of the deceased may be present. These places require a different order and ways of cleaning them than available within a Christian set of practices. The economic logic behind these places is one of gifting rather than selling. The community cannot use *agbadome* places for other purposes whilst they are inhabited by spirits, which, as I will show later, may call for an intentional soiling of the areas in order to transform them from spirit places to worldly places.

With Douglas, cleaning is a way of reinstating social order in the face of uncertainty. The handling of dirt, exactly because it is the source of potential danger, is important for patching up social structures that are broken down in death. Douglas does not explore the material qualities of dirt as matter. Dirt, however, is manifested as matter after all. In this sense, an approach which considers the intersection of cleaning practices and practices around death in relation to the material can prove fruitful for understanding processes of negotiating power in a society or a community.

In the town, the public presence of dead bodies stands at the center of these cleaning practices. The washing of bodies is carried out in a particularly public setting amongst a crowd of mourners, undertakers and relatives at the morgue. In this process, the use of specific toiletry objects and the status of the morgue as a state-owned institution are both equally important for negotiating the deceased's social value within the community. Only a positive evaluation and the settling of possible debt to the community through the family – be it monetary or moral – can make a body eligible to be washed and buried.

Cleaning the Body

Unlike most towns in the vicinity, the town has its own morgue. It is state property and part of the local 'Peki Government Hospital' complex. The morgue does not only store the bodies of town residents but also those of people from the wider area. At any time of year, at least several dozen bodies are waiting in deep freeze at the morgue's cold room.[3] With such a large quantity of bodies in the small confines of the morgue it gets crowded easily. On Fridays in particular, the morgue is full of people attending the washing of a relative. The two washing tables are in constant use and, for lack of space, undertakers continue to work on the bodies on the floor after they have been washed. Nothing here is private: visitors are part of an incidental public and inevitably witness not only the washing of their relatives but of other bodies as well.

Each family delegation arrives with an undertaker and one or two designated *tovi*. The *tɔvi* – a symbolic role given to a family member as custodian of the person in life and death – have to accompany the body of the deceased during the funeral proceedings and they have the duty, among other tasks, of bringing a bucket with cleaning products and underwear to the washing. The undertaker carries out the washing while the *tɔvi* provide the tools. The bucket, a gift to the deceased, is supposed to be a new plastic bucket which usually contains a bar of soap, a sponge that is made out of a synthetic fabric mesh, white talcum powder, a new towel and a set of new white underwear packaged in plastic (fig.7). For this occasion, the properly packed washing bucket contains things that are new and marked as purchased goods, for example by their branding or wrapping. These commodities are treated as more valuable and more appropriate for the occasion than were the bucket to contain used, organic or unpacked washing items.

3 Rural electrification has made the maintenance of morgues possible in Ghana. Yet, as much as people rely on these morgues, electricity cannot be guaranteed at all times. In the town, the hospital and morgue complex have their own generator, securing continued light and deep-freeze for both even at times when residents of the town are without electricity.

Figure 6. Bucket with gifts for a deceased used during a washing at the morgue. (Photo: Isabel Bredenbröker).

A popular personal hygiene item for Ghanaians is the net sponge, a net-like fabric mesh made of cotton or nylon that is cut to size but usually quite long. Estimates suggest the net sponge has been in use in West Africa since the 1950s. Their organic predecessors are loofah sponges or loose coconut or palm tree fibers that are scrounged together. In a market, I come across a finer and a thicker version of natural fiber sponges. The vendor tells me the finer version is reserved for the dead. However, I have not seen this type of organic fiber sponge in use on a dead body. Instead, the bodies of the deceased that I see at the morgue are washed like those of a living person. New, plastic-wrapped and packaged objects with a preference for non-organic materials are carefully chosen for direct contact with the body. It is the responsibility of family members to provide these things and by selecting the more prestigious options of new, clean things, the act of washing the body

Figure 7. Sponges and soap bars on a window at the morgue. These are used for washings of bodies that are not washed and gifted by their relatives. (Photo: Isabel Bredenbröker).

itself is given elevated importance. Family obligations and the materials used to carry them out are contextualized as a joint effort between a state-owned site, locally resident workers supervising the site, a professional undertaker and family representatives to establish a new sense of personhood of the deceased in death.

For the washing, the body is laid on one of two washing tables in the morgue. From a big plastic container, the undertaker ladles water with a calabash that floats on its surface.

The body is covered in soap which is foamed up with the sponge. Just like a living person would do for themselves, all parts of the body are scrubbed rigorously, including the face. After washing everything off, the body is dried with a towel, the hair is combed into shape, the undertaker rubs antibacterial sanitizer and perfume into the skin and the body is then lifted onto a mat for further dressing and cosmetic interventions. The *tɔvi* can actively take part in this washing, but mostly do so only symbolically by ladling one serving of water over the body at the beginning. During washings that I attend, bodies are dressed in new underwear in the morgue. Until bodies are picked up and up until they reach the house, the need to be contained in a white body bag, as prescribed by the morgue regulations. The morgue also sells this body bag at a fixed, quite high price of 50 Cedi. The body is then either wrapped in a piece of Kente fabric and dressed later in the house, or it is dressed at the spot in several layers of their own clothes. When laying the body in state, a costume for the occasion such as a suit with an open back or a satin and lace dress is added as the last layer. These may be made specifically for corpses or bought new to be used as a dress for a corpse, cutting the back open to make dressing easier.

The sponge and soap used during the washing are usually included in the coffin so the deceased will be able to wash in the beyond. The act of washing is thereby given a continued importance. The bucket and some of its contents are supposed to be incorporated into the *tɔvi's* household after the washing and are therefore intended for use by the living. Such a reincorporation of material things used to clean the dead body is a symbolic gesture of maintaining contact to the deceased. It implies that these objects hold no potential danger, are not dirty. However, this is only performed to a certain extent these days. People voice concerns over hygiene issues and mostly, so I am told, it is only the bucket that ends up being re-used.[4] Undertakers bring their own equipment, which, as work tools, are used and re-used indiscriminately. Economic activity does not prescribe the same 'newness' of materials as does the act of gifting the bucket. Yet, the kinds of materials which are used are similar, for example commodity cleaning products, synthetic fabrics and fashion jewelry. Marcia[5], an undertaker who proudly shows me her collection of decorative props and work utensils, has her own sanitizers and perfumes that she mixes at home and brings to the morgue for washings. Florida Water for example, an old-fashioned cologne, is used as a popular scent for rubbing the skin. Marcia brings make-up and materials like cotton wool and super glue as part of her kit. All objects and fabrics used on and around the body at the laying-in-state are constantly cleaned and re-used.

Marcia's work at the morgue depends on the goodwill of the staff, who, as representatives of the state, have the final say as to who may use the morgue as a workplace. As a state-owned institution, the morgue is a site where the state's regulations regarding cleanliness and proper conduct are the ground rules, including set times and payments. The mandatory white body bag comes at a quite high price considering that it is only used while in the morgue and during a twenty-minute trip into town when the body is transported. There, the body is unwrapped by the undertaker and presented to visitors who may interact with it and touch it without the interference of the state in the form of a protective layer such as the body bag. At the morgue, the state's presence can be noted in

4 The bucket does not come into contact with water but only serves as a bag for the toiletries and other washing implements.

5 No real names are used in this paper to protect individuals

the form of a member of the Public Health Office who is supposed to be present during the Friday washings. The director of the morgue controls the execution of these rules. Her role in relation to the state is an indirect one however, since she does not have a permanent work contract which would provide her with a state pension. She enters all payments, bodies and pickup dates in a large accounting book and ensures that everything is duly registered. In cases where a body cannot be washed by an undertaker and relatives, her team of two assistants perform the washing, quick and unceremoniously, without the valuable presence of family and the carefully chosen objects in the bucket. Their washing tools, a few voluminous soft foam sponges and used soap bars, lie on the open window glass panes next to one of the washing tables and are used over a longer period of time (fig.6). This way of cleaning is a mere act of mechanical duty, not a ritual with multiple participants which honors the personhood of the deceased. Unaccompanied bodies are subject to the state's set of rules and regulations. Some bodies are never claimed back for washing and funeral. In these cases, the bodies remain at the morgue and are subject to the evaluation of the director. The corpse of a murderer who was killed in an act of public vengeance in town is merely tolerated. The director does not consider it worthy to stitch up its injuries, let alone washing it or organizing a burial. In such cases, the personhood that is re-instated into the corpse through washing and further rites is denied to the deceased. Suicide victims are treated in a similar way. While they are admittedly buried, they are neither washed nor laid in state but transported directly to the cemetery. In an attempt to scare others off from such morally bad behavior, bodies may be handcuffed and all activities that show last respects to the person have to happen on the way to the cemetery, following the coffin.

As becomes apparent, Friday washings at the morgue bring people working as state authorities, private entrepreneurs, families and mourners together in an intimate yet public setting. Rules are put forward by the hospital authorities in representation of the state but practices as performed by all participants, as well as the materiality of bodies and of cleaning items create a reality different to those by-the-book regulations. The limitations of the space, an overflow of bodies alive and dead, cleaning materials, mourners, symbolic roles taken on by kin, paid work, state-imposed rules and the ways in which these rules are or are not enforced, shape an understanding of cleanliness in the process.

Cleaning, Social Organization and Power

Just like washings taking place at the morgue, other public cleaning processes equally involve community members, community rule-makers, public and private sites as well as evaluation processes. As we will see, the ways in which different sorts of public cleanings are organized directly feeds back into cleaning processes in relation to death. Looking at the ways in which public cleaning is administered and how it is tied to obligations and evaluations helps to understand 'the multiplicity of relations to power' at play here (Duschinsky, 2016). Who defines the rules of cleanliness and who is involved in doing the cleaning? What are the stakes?

Unpaid community labor is an essential part of community organization and is scheduled by the Palace Staff of local town chiefs. Tasks include filling in potholes and cutting back overgrowth along roadsides, building public toilets and town cleaning. These tasks are made public by morning loudspeaker announcements. The cleaning and care for sites related to death are partially included in the community labor tasks, hereby making a cemetery a place that the community is responsible for. Participation in community labor is mandatory in order to be deemed a good community member. The outcome of such an evaluation – something which certainly does not stand undisputed and may look different for individuals depending on who evaluates them – plays out in multiple ways. This complex combination of different views may be motivated by intentions like increased profit, clan and family ties, claims to a political position or making someone socially unacceptable, for example as an act of revenge. Members of the community who permanently reside in town pay a low annual contribution of five Cedis into the community account. Absent community members are charged a higher fee which they must pay in order to compensate for their absence during community labor. If people fail to perform the role of a good and active community member by participation in cleaning and labor (or through payments that stand in for these activities) their family will be fined. The severity of the fine(s) can vary, and claims can come from several parties in the aftermath of a death, for example from within the immediate and wider family, from other families or from the chief's palace as well as other local bodies such as associations. The unpredictability of such claims makes their looming possibility unsettling for relatives of the deceased. It also shows how multiple 'relations to power' (Duschinsky, 2016) can become relevant and manifest themselves in financial claims which also function as moral judgements. Until the payment has been agreed on and settled, the family is not allowed to wash and bury the body. In that respect, the deceased and their family have become socially unacceptable, the body remains morally stained, unwashed and, to a degree, 'out of place'. These negotiations take place on a community level and consider factors such as the (not always realistic) estimation of a person's economic situation and their general conduct of life. Only a clean person, meaning a person whose family members, local authorities and possibly state representatives make no further claims, is allowed to be washed and properly buried and thereby put in the right place. Monetary payments are often used to settle these claims but, as we will see later, money is not always the appropriate medium to clean up after the dead.

The state participates in this process of evaluating the moral personhood of community members in the form of the local hygiene patrol and in relation to home burial regulations. In both regards, it holds a powerful position as a controlling and economic force and can reprimand citizens for 'unhygienic' conduct. With the purported intention of educating people about the proper maintenance of toilet facilities, rubbish disposal and water storage in the house, a team from the Environmental Health Office (EHO) performs unannounced neighborhood visits, dropping in at people's houses and thus traversing domestic and public spaces alike. While 'education' sounds well-meaning, the practice presents itself somewhat differently. When a household is found non-compliant with sanitary standards, they may be subject to fines, warning notes on their properties or, if the problem is not settled, a hearing in court. The EHO also controls the formerly very popular (and more prestigious) practice of home burials which was recently banned by the state for hygienic reasons. If a home burial is requested, this may be allowed under supervision by a EHO representative for a very high payment of 5000 Cedi.[6] This way, the state has monopolized

6 5000 Ghana Cedi is equivalent to about 1000 Euros. This is an extraordinarily high sum – absolutely unaffordable for most people in the town. In cases of low-expenditure funerals, the total sum of expenses would be be well below that sum.

the most prestigious form of burial and made it legally unavailable for most people. A new 'modern' private cemetery recently opened up in town, challenging this monopoly. The idea of the cemetery owner was to give his graves the same price tag as a home burial and offer a better location than the public cemeteries. If one is financially able to bury here, all demands for payments which are based on one's performance as a community member fall flat. The cemetery offers a way out of negotiating one's social standing.

Sites, Materials and Practices

There are currently eight cemeteries in town, all subject to rules made by different authorities. Cemeteries may be owned by the community and supervised by the local chief's palace, or by individuals who are loaning the land to the palace. The palace usually appoints an overseer as a volunteer to collect payments for graves and appoint plots. There is one cemetery that is mainly intended for members of several clans as well as one that is privately owned and charges for burials as a private business. The maintenance and cleaning of these cemeteries is either partially a communal job, partially a job of the relatives, family or clan of a deceased, or it can be done professionally. This will depend on the way the cemetery is organized, on the social status of a deceased person and their family, as well as on the ability to purchase a fully serviced grave in the private cemetery.

One of the largest cemeteries in town, for example, is under custody of the local chief and an appointed cemetery overseer. Payments for grave plots in communal cemeteries are made into a local community account which is under administration of the chief. The overseer lets people know where their plot is and shows up on Wednesdays when graves are being dug by young men from the clan of the deceased. Three different kinds of graves are on offer. Their materiality and the payments attached to them directly reflect the social status of a person. The cheapest option is a simple pit decorated with moveable or impermanent things like a metal sign, plants and a wreath. Muslim graves, as exceptions, are priced significantly higher than the plain graves whilst also being located outside of the cemetery and merely decorated by piles of stones. This indicates that members of the Muslim community are to some degree treated as second-rate citizens by town authorities, in this case by the Chief's palace who is in charge of cemetery organization. For five times the price of a plain grave, one may buy a plot that then gets cemented inside and out, practically creating a cement grave chamber underground and a solid structure above ground, both of which are almost impossible to remove. The cemetery creates the major part of the community budget, money that is guarded by chiefs and used for communal expenses such as the building of toilets. In terms of cleaning up and re-using the cemetery, the official statement I get which explains the difference in price between impermanent graves, cemented graves and home burials is that higher payments are supposed to discourage practices which make graves permanent installations and land re-use impossible. These graves cannot be cleared away. As a result, more expensive graves (with further expenses for decoration) will last and so will the memory of the people who are buried within them. Monetary payments, social valuation and the materiality of graves thus determine the kind of afterlife a deceased has in the community.

After a funeral, the cemented graves turn into construction sites, depending again on the family's ability to spend money for commemorative and funerary purposes, and are constantly worked on until they are unveiled in their finished state a year after the funeral. The deceased is usually believed to enter the realm of their ancestors on that day. Popular materials for decorating the permanent graves are cement as a base and upon this either a structure that is covered in shiny tiles, decorated with durable terrazzo or, in its most expensive version, made from imported granite with an engraved text on it. In comparison to the impermanent graves, these materials are more expensive than living cassava plants, commemorative objects such as lorry tires for a truck driver or a sewing machine for a seamstress, and metal signs and wreaths. The wreaths, which also feature prominently on the cemented graves, are structures made out of cardboard, gift ribbon, colourful cellophane foil and synthetic flowers, wrapped in transparent foil. These wreaths remain wrapped even when placed on the graves and are small, shiny and colourful units of synthetic materials throughout the cemetery. When compared to their organic counterparts – wreaths made out of palm tree branches – locals value the synthetic versions as more appropriate due to their bright colours and resistance to decomposition. Overall, imported or synthetic materials of different kinds are valued as good for use in relation to death, be it for washing the body, decorating and installing the grave or as donations to the deceased. The use of such materials by families and individuals in the context of funerals and beyond can express a degree of prestige as well as a relative's willingness to give the deceased things that will last. Once installed, the grave does not need a lot of maintenance but might eventually receive an update when another relative is buried and an order for a carved stone is made.

Apart from the cemeteries, there are also three sites in town that serve as donation grounds to the spirits of people who have died suddenly. These sites, *agbadome*, meaning center of the clay bowl in Ewe, are used to deposit luggage for these spirits and are also considered dwelling places of these spirits. One *agbadome* site next to one of the larger cemeteries is located in a wooded area. Donations of plastic chairs, tables, suitcases and bags packed with clothing, pots, cutlery, jars and toiletry items are left in the woods. This site is not supposed to be entered without pouring libation and asking permission. Spirits of the deceased who died without warning, for example accident victims, are said to live here. Depositing things at this site keeps spirits pacified and uninterested in the world of the living, preventing harm to other family members. A small path demarcates the borders of the site. Another *agbadome* site in a different community of the town has started to overlap with a rubbish ground nearby. At the same time, the rubbish is being used for ritual purposes associated with the spirits, mixing the sacred with the profane. Bad spirits, so I am told, feast on rubbish and can thus be found there. A bad dream might call for a visit to the rubbish dump to address the issue in conversation with the negative spirits. A thin makeshift piece of red thread now marks the border between this dump and *agbadome*. Materially, it is hard to tell one apart from the other. A Christian undertaker tells me that she sometimes deposits luggage for people at the rubbish dump instead of *agbadome*, not wanting to come into contact with the spirits there but instead using a less spiritual place that might serve the same means. At the same time, I learn that there are plans to build a new palace for the local chief on the site that is now a dump / *agbadome*. The blurring of categories between different kinds of spirits, donations and rubbish would then lead to a possible reusability of the land, contrary to the impact of cemented graves in a regular cemetery..

In the cemetery, acts of not cleaning are also relevant to express dissatisfaction about an abuse of power. When the chief in charge of the community payments for this cemetery fails to give account of how the money has been used, so I am told, the women whose communal job it is to keep the cemetery clean and cut back the bush go on strike, protesting a possible misuse of the communal funds. Not cleaning, or instead leaving the maintenance tasks to families in relation to their relative's graves, thereby becomes a form of protest regarding the use of communal money or its abuse by authorities. The cemetery becomes a site for enacting this.

Conclusion – Social Organization and Power Struggles

Practices related to death and cleaning practices partially overlap and form a nexus which connects actors and actions from the cosmic sphere, the state, the family and local authorities. These spheres interact with one another in the process of cleaning or refusing to do so. The aim of cleaning or attributing 'cleanliness' in death is the production of a socially 'good' death. This happens in a communal process of evaluating a person's contribution to the community and possibly making up for defaults through payments. It also requires the representation of the dead within a continuous present time: stable, continuous and 'alive', albeit in a different sense.

New Materialist ideas of 'vibrant matter' (Bennett, 2010) – matter alive which transforms continuously – can help us to understand material representations of death (and their limits) as a vital state. In this ethnographic context, potential properties that imply a higher likelihood of change in matter are only partially favored in materials around death. Synthetic materials are predominantly evaluated as the better representation for this state of permanence and aliveness than their organic counterparts are. Nevertheless, organic materials, such as soil in representation for an accident victim's soul, are also used with the same implications. The choice of materials and practices deployed in relation to a deceased – at a lying in state, a washing and in the construction of their grave – are crucial in the making of positively appraised, post-mortem personhood. This marks the culmination of an evaluation process which occurs throughout a lifetime and, in death, includes valuations motivated by different agendas of the living which may be unconnected from the actual deeds and failings of a deceased.

On a social level, multiple agendas are at play in the process of establishing a deceased's 'clean' status as a person who can be buried without objections. Claims to power – be it economic, political or spiritual – are directly related to the engagement of different parties and things in processes of cleaning. State powers such as the cleaning patrol, laws, fines and payments, as well as oversight of institutions like the morgue, play a significant role in controlling cleaning processes and funerary proceedings. Checks like those of the cleaning patrol which seem unrelated to death directly influence the social evaluation of citizens as community members. The conditions under which one becomes a socially respected person are related to active engagement in the community, good moral conduct and general cleanliness, including the cleaning of the community. If these things apply, a person can be considered morally clean. But rather than this being a clearly identifiable condition, it is produced through negotiations between different parties until a settlement of demands is reached. In this sense, payments which are often requested to make up for a lack of social credit also have a cleaning function, for the deceased as well as for their relatives. In death, the judgement of a person's social value can be finalized and have persistent relevance for the living.

The conditions under which one may die a good death are largely out of an individual's control and, if the death happens to be 'bad', the need to produce a state of cleanliness, spiritually and physically, has to be shouldered by the immediate, nuclear family.[7] The *agbadome* sites are places where spirits reside in states of limbo. Dirt and waste are only a small step away from belonging to another spiritual world and do in fact take on spiritual meaning, thereby having a sacred function and acting very much as matter 'in place', subsequently leading to a re-interpretation of these sites which, as rubbish dumps, are considered fit to be used for building. Practices that deal with death – as well as cleaning practices – are arenas in which multiple rules are made and contested. These rules have been made for economic ends and on a local political level, based on spiritual and moral grounds.

Bibliography

Bennett, J. (2010) *Vibrant Matter. A political ecology of things*. Durham, Duke University Press.

Bernault, F. (2006) Body, Power and Sacrifice in Equatorial Africa. *The Journal of African History*. [Online] 47 (2), 207-239. Available from: http://www.jstor.org/stable/4100722 [Accessed 14th May 2018].

Bloch, M. & Parry, J. (eds.) (1982) *Death & the regeneration of life*. Cambridge, Cambridge University Press.

Braidotti, R. (2010) The Politics of "Life Itself" and New Ways of Dying. In: Coole, D. H. & Frost, S. (eds) *New Materialisms. Ontology, Agency, and Politics*. Durham, NC, Duke University Press, pp. 201-218.

Coole, D. H. & Frost, S. (eds.) (2010) *New Materialisms. Ontology, Agency, and Politics*. Durham, NC, Duke University Press.

Douglas, M. (1966) *Purity and Danger. An analysis of concepts of pollution and taboo*. London, Routledge & Kegan Paul.

Drazin, A. (2015) To live in a materials world. In: Drazin, A. & Küchler, S. (eds.) *The Social Life of Materials. Anthropological Accounts of Materials and Society*. London, Bloomsbury, pp. 3-28.

Duschinsky, R. (2011) Purity, power and cruelty. *Critique of Anthropology*, 31 (4), 312-328.

Duschinsky, R. (2013) The politics of purity. When, actually, is dirt matter out of place?. *Thesis Eleven*, 119 (1), pp. 63-77.

Duschinsky, R. (2016) Introduction. In: Duschinsky, R., Schnall, S. & Weiss, D. H. (eds.) *Purity and Danger Now. New Perspectives*. London, Routledge, pp. 1-19.

Duschinsky, R., Schnall, S. & Weiss, D. H. (eds.) (2016) *Purity and Danger Now. New Perspectives*. London, Routledge.

Hertz, R. (ed.) (2004) *Death and the right hand*. 2nd ed. Routledge Library Editions. Anthropology and Ethnography. Abingdon, Oxon, Routledge.

Huntington, R. & Metcalf, P. (eds.) (1979) *Celebrations of death. The anthropology of mortuary ritual*. Cambridge, Cambridge University Press.

Jindra, M. & Noret, J. (eds.) (2013) *Funerals in Africa. Explorations of a Social Phenomenon*. 2nd ed. New York, Berghahn Books.

Kim, H. (2012) The Purification Process of Death. Mortuary Rites in a Japanese Rural Town. *Asian Ethnology*, 71 (2), 225-257.

---

7 Bad death being any death of a sudden or violent nature, such as accidents, murder, suicide or unexpected termination of life for unknown reasons.

Lamont, M. (2013) Decomposing Pollution?. Corpses, Burials, and Affliction among the Meru of Central Kenya. In: Jindra, M. & Noret, J. (eds.) *Funerals in Africa. Explorations of a Social Phenomenon.* 2nd ed. New York, Berghahn Books, pp. 88-108.

Locke, T. A. (2016) *The Living and The Dead. An Essay on the Tendential Continuities of Life and Death.* London, Repeater Books.

Lopez, P. J. & Gillespie, K. A. (eds.) (2015) *Economies of Death. Economic logics of killable life and grievable death.* Abingdon, Oxon, Routledge.

Meyer, B. (1999) *Translating the devil. Religion and modernity among the Ewe in Ghana.* Trenton, NJ, Africa World Press.

van Gennep, A. (1909) *Les rites de passage. Étude systématique des rites de la porte et du seuil, de l'hospitalité, de l'adoption, de la grossesse et de l'accouchement, de la naissance, de l'enfance, de la puberté, de l'initiation, de l'ordination, du couronnement, des fiançailles et du mariage, des funérailles, des saisons, etc.* Paris, Émile Nourry.

van Gennep, A. (1977) *The rites of passage.* 3rd ed. London, Routledge & Kegan Paul.

The Cultural Aspect of Cleaning in Archaeology – a Case Study from the late Neolithic Site of Fıstıklı Höyük

Georg Cyrus

Introduction

The categorization of whether something is clean or unclean involves constant negotiation. Most people who live together will sooner or later experience that the value of cleanliness is to a large part subjective. Depending on context and socialization, there are different views on how to clean and the satisfactory state of cleanliness. For instance, people would generally agree that cleaning plays an important role in hospitals; but the same people might disagree on how to clean the kitchen or bathroom. In this paper, I will reconstruct uncleanness and cleaning activities in the late Neolithic site of Fıstıklı Höyük in upper Mesopotamia, a small and short-lived Halafian settlement inhabited by partially mobile communities (Bernbeck & Pollock, 2003; Bernbeck, 2008). I will accordingly focus on the value of cleaning at this site by considering clean and unclean as socio-culturally constructed categories. Today, there seems to be a general consensus that unclean conditions result in health issues or, ultimately, even in death. Nevertheless, every culture deals with these matters in different ways, and I attempt to investigate these culturally specific activities of Fıstıklı Höyük, rather than their biological causalities.

Cultural activities are extremely diverse and, even within one society, values and activities around cleaning and uncleanness differ from context to context and group to group. Hence, I assume that past societies were similarly complex in their views on what was to be regarded as clean or unclean. However, it would be a mistake to transfer our own contemporary Western view on (un)cleanness onto other cultures, be they past or present. Whilst engaging with other cultures, the archaeological concepts of *emic* and *etic* are helpful in describing two alternative perspectives on a culture. Whereas emic describes the view from within the culture, etic refers to the observation from outside (Pike, 1967:p.37; Harris, 1979:pp.32-41).

In archaeology this distinction has been used to classify different research interests, especially between processual archaeology, interested in an etic perspective, and post-processual archaeology that researches according to the emic view (Bernbeck, 1997:p.46).

in: Bredenbröker, I., Hanzen, C. & Kotzur, F (eds) 2020: *Cleaning and Value. Interdisciplinary Investigations*, Sidestone Press (Leiden), pp. 89-104.

The two terms, first coined by Kenneth L. Pike (1967), are based on the linguistic terms phon*emic* and phon*etic*. Both are classifications of sounds, but while phonetic classifies a sound through the function of the body, phonemes are the units of a specific language and therefore are distinguished by their significance for that specific language. Pike (1967) defines the terms emic and etic analogous to their linguistic usage, 'but for more general purposes' (p.37). He is interested in analyzing cultures, and thus 'etic' pertains to the objective scientific view from outside, while resorting to 'emic' to describe the internal logic of the analyzed culture. Marvin Harris (1979) transfers this model to the ethnological field with the aim to conceptualize the relationship of the mental and the behavioral elements of culture (pp.32-41).

At its core the emic/etic pair remains still a linguistic concept. It can therefore be connected up with the theoretical work of Mary Douglas which is also based on a linguistic approach to culture.

My aim in this paper is to show the feasibility of this dichotomy in constructing a middle-range theory. Therefore I will first introduce etic categories to prove hypothesis concerning emic view with a middle range theory to subsequently arrive at a better understanding of the emic categories. In doing so, I will attempt to address the questions of whether and how we can reconstruct concepts of uncleanness in prehistoric societies.

Firstly, I will define the category of uncleanness and the process of cleaning using the work of Mary Douglas (1994). I will then introduce and evaluate the sociological work of Norbert Elias (1939). After that, I will move to an archaeological discussion on uncleanness, where I focus mainly on the methodological works of processual and behavioral archaeology (Schiffer, 1972; Schiffer, 1976) and reflect on them in relation to post-processual critiques (Moore, 1982). To tackle the challenge of reconstructing emic categories, I will present the method of micro debris analysis and will formulate testable hypotheses. In order to show this challenge in practice, I will present as a case study the site of Fıstıklı Höyük, Şanlıurfa province, Turkey.

Sociological Theories

If I want to discuss what is unclean in other cultures, I first have to transform uncleanness from a common word into an analytical concept. Where common words are mostly vague and open to interpretation, analytical concepts have to be specific and explanatory. Analytical concepts do not have to be universal categories but must nonetheless hold validity for more than one single case (Bal, 2011:pp.13-15). To conceptualize the word uncleanness, I draw on the conclusions of Mary Douglas.

Douglas' work *Purity and Danger* can be read in several different ways. Here, I will present a reading largely focused on categorization as the inherent human ability to create culture. Douglas (1994) investigates and compares different religious concepts of impurity such as ritual uncleanness or dietary laws and concludes a universal definition for what she interchangeably labels pollution, dirt and uncleanness. In this paper, I use the term uncleanness with reference to Douglas' definition. In the process of working out this definition, she examines a series of different cultures, including Western European ones. She argues that different definitions of uncleanness are historically and culturally specific. With the model of 'medical materialism', she describes and historicizes the Western definition of Uncleanness and poses the question whether all ideas of uncleanness have something in common (Douglas, 1994:pp.30-41). According to Douglas, the medical ideal of purity is deeply rooted in the medieval fear of pandemics and the modern discovery of bacteria. A Paleolithic society had no experience of pandemics (Wolfe, Panosian and Diamond 2007, Scott 2017) and was not aware of micro-organisms. Therefore, their ideas around cleanliness must have been based on other socio-cultural factors. For that reason, Douglas' main question is whether all ideas of uncleanness have something in common.

Mary Douglas starts by defining the human ability to make culture through language. Every human categorizes the world around him or her and these (more or less strict) categories form the social reality that people live within. Douglas defines 'unclean' as everything that is not categorizable; it could thereby be viewed as the typical leftover category 'other'. The non-categorizable refers to a known category in the wrong context, for example, shoes on the breakfast table. Shoes alone are not unclean, only at the moment they cross the borders of contexts. This definition of uncleanness as border-crossing is the plausible basis for reconstructing emic categories.

Douglas' model historicizes concepts of uncleanness and illustrates that they are socio-cultural products. However, she does not consider the dimensions of power and violence that are present in most societies. Power enters the concept of uncleanness in diverse forms. One example that might be visible in an archaeological context is disciplinary power. As Foucault's analysis of modern societies shows, discipline is a particular form of power widespread in modern nation states with institutions such as prisons, barracks and schools (1977). How discipline and the concept of uncleanness are connected is best stated in the research of Norbert Elias.

Elias' main work *Über den Zivilisationsprozess* is a long-term study of the transformation from medieval into early modern European society (Elias, 1939). Elias uses a variety of different sources to analyze this transformation, describing its historical process as the change from a society controlled by direct violence in the early medieval period to one governed by indirect violence in the modern era. Over time, monarchs progressively monopolized violence and people became dependent on a centralized economic system. To deal with this new state of dependence, self-discipline and future-planning became necessary. Indirect violence and such forms of discipline have been identified as causes for increased psychological problems in Modern societies. For Elias, the civilizing process is therefore an ambivalent one, and he accordingly questions the simple narratives of progress that dominated his times. For Elias, this ambivalence is also reflected in the quality of self-discipline and the activity of future planning, both of which are commonly viewed as positive. Going forward, I will use the term self-discipline as understood by Elias, meaning an internalized form of control and of violent disciplinary measures that is ambivalent in its output.

Elias identifies cleaning activities as a main index for the changing mindset of European society. His core idea is that fear of breaking rules goes hand in hand with internalized and increased self-discipline. He conceptualizes this fear with the help of the two terms *Scham* (shame) and *Peinlichkeit* (embarrassment). According to Elias, *Scham* is the feeling experienced by a person breaking social rules, whilst *Peinlichkeit* is experienced when someone else breaks these rules. These two feelings increase throughout the process of civilization and they lead to an increase in self-coercion. For Elias, the connection of self-discipline and fear is materialized in cleaning activities.

Because of the fear of *Scham* and *Peinlichkeit*, people tend to control their own and others' uncleanness more strictly. It is possible to connect the respective models of Douglas and Elias: when indirect violence and interdependency increase, self-discipline and fear grow, and categorizations become stricter. In other words, a society that exists with less internalized pressure resulting in the ambivalent form of self-discipline will have more open emic categories. This consideration has clear implications with regards to (un)cleanness. Because Douglas defines uncleanness as a transgressive state, a less disciplined society could be expected to evaluate less things as unclean, than a strongly disciplined one. At this juncture, I can start thinking about the archaeological record which often also results from cleaning activities. These cleaning activities too can be investigated and evaluated as construction and retaining of the emic categories. In order to do so, however, I need a feasible method.

Archaeological Approaches

Refuse, according to our current understanding of the term, is a common feature of archaeological records. Garbage pits filled with bones and pottery or refuse gathering in abandoned houses constitute some of the main sources for the reconstruction of the past. In this article, I use the word refuse as the loose category in archaeology to describe finds and contexts which we would today consider unclean. The meaning of the word 'refuse' is rooted in our cultural practices – we describe broken pottery, production trash and food remains as refuse, whereas in other cultures these objects might be viewed as recyclable goods. Nevertheless, refuse is also an analytical concept for archaeologists that can help in the understanding of uncleanness and cleaning activities. In other words, while dirt is the universal leftover category characterized by border-crossing, refuse is the specific kind of uncleanness conceptualized in the archaeological discussion as a specific type of deposit.

Refuse was already recognized as an important archaeological source in the nineteenth century. A famous example is the Danish Køkkenmøddingern (Trigger, 1989:p.82), or later, the shell mounds along the west coast of North America (Gifford, 1916). Danish archaeologists interpreted these shell mounds as kitchen refuse and reconstructed past lives through inductive reasoning.

Systematic research on the potential and limitations of refuse in the archaeological record was first conducted by Michael B. Schiffer in Behavioral Archaeology (Schiffer, 1972; Schiffer, 1976). He aimed to reconstruct the formation processes that lead from a systemic context[1] to an archaeological one. To do so, he categorized objects according to deposit type. He identifies three types of refuse: primary, secondary, and de facto refuse. Primary refuse is 'the material discarded at the location of use' (Schiffer, 1972:p.161). Secondary refuse is relocated primary refuse that no longer remains at the location of use. Finally, de facto refuse is refuse that has been abandoned. Technically, it is not therefore refuse, because it remains in its position, but it is de facto, because it enters the archaeological context. For the archaeological definition of unclean and the activity of cleaning, notions of primary and secondary refuse are useful since, for Schiffer, cleaning is the transformation from primary to secondary refuse.

1 The systemic context is the living society that we want to reconstruct. The word emphasizes the typical processual definition of culture as system (See Binford, 1962).

Establishing the most likely hypotheses to explain human behavior was one of the programmatic goals of processual archaeology (Binford, 1962). With the classification of primary and secondary refuse, ethnoarchaeological projects tried to establish hypotheses to explain human cleaning behavior. Observation of living societies has the advantage that the removal of unclean things, as well as the values for doing so, can be recognized and correlated. An early ethnoarchaeological work that tried to establish universal human cleaning behavior is the comparative study by Priscilla Murray (1980). She compared seventy-nine ethnographies and established the hypothesis that humans always clean their living space.

Direct fieldwork with living communities leads to much more complex observations. Most hypotheses were proven wrong due to the sheer complexity of these societies. The famous example of the Coxoh Ethnoarchaeological Project ran from 1977 to 1979 and took a so-called 'skeptical processual' approach (Hayden & Cannon, 1983). The complexity of Maya households led the team to abandon universalisms and instead observe trends. They noticed that the residents did not treat all refuse equally. Rather, residents were driven by an 'economy of effort', meaning efficiency motivated them. Hindrance potential and reuse value were the basis for their decisions. Different spaces were cleaned differently, but so-called 'randomizing forces' like children's play or animals obscured patterns. The scholars had a realistic viewpoint on the complexity of cleaning activities. On a methodological level, they observed that it is more meaningful to take the whole settlement and not the household as a unit of investigation – an observation that excavators also made in Fıstıklı Höyük where it is rather difficult to identify economic units like households.

One of the first ethnoarchaeological works that specifically focuses on the symbolic dimension of material culture, and therefore stands at the beginning of post-processual archaeology, was the study Henrietta Moore (1982) conducted in the 1970s. While investigating refuse patterns of the Marakwet in Kenya, Moore observed a link between genders and refuse patterns. The society of the Marakwet is sedentary village based, with a strong binary gender system that manifests itself in architecture, production activities and even in the refuse resulting from production. Men are responsible for animal husbandry and the related refuse belongs to the male sphere. In contrast, women are responsible for cooking activities and therefore ash and other related refuse belong to the female sphere. Male refuse has to be located behind the men's huts and female refuse behind the women's. This is a good illustration of how emic categorization is reflected in the find distribution which cannot be explained by a functionalistic approach.

The ethnographies of the Maya and the Marakwet try to correlate the emic view of (un) cleanness with the activity that leads to specific archaeological patterns. Obviously, the values of those like the Marakwet are difficult, if not impossible, to reconstruct from the archaeological record alone. Nevertheless, this case has implications for archaeological research and can be used as analogy, inspiration or critique for the interpretation of the archaeological record. It is now necessary to take a closer look at the archaeological sources to identify the potential this source has in light of the ethnographic examples, the archaeological concept of refuse and theoretical approaches to uncleanness.

Methods and middle-range Theory

On the basis of Douglas' definition of dirt and Schiffer's concepts of primary and secondary refuse, as well as keeping in mind the examples of the Maya and the Marakwet, I can formulate a few hypotheses that are testable with the archaeological record. They do not directly reveal the emic categories, but I argue that they are an approximation towards these categories.

The observable information in the archaeological record consists of materialized activities from past subjects. To formulate premises means, in this context, to connect specific activities with such observable material patterns. I will therefore express hypotheses that are based on the methods and theories presented and can be tested with archaeological evidence. Through the reformulation of these assumptions, I can establish a plausible hypothesis about the emic categories that were materialized through cleaning activities.

Following Douglas in her definition of uncleanness, people experience the unclean as something in the wrong place which is disturbing emic categories. The most plausible activity connected to this experience is cleaning; to put that which is unclean in the place where it belongs and thus to restore emic categories in the material world. In terms of behavioral archaeology, cleaning is the transformation from a primary refuse to a secondary refuse, and I can observe the result of this process in the archaeological context with micro-debris analysis.

Micro-debris analysis is based on Diana Giffords' (1978) investigation of the correlation between the size of an object and the probability of cleaning it away. She observes that the smaller the object, the more likely it remains in place despite cleaning efforts. Moreover, if the surface is made out of a permeable substrate like clay, smaller objects likely become trampled into the subsurface zone. These micro-finds then become primary refuse that cannot easily be cleaned away. Investigations into micro-debris are already known from several activity area studies (first Fladmark, 1982; later for Southwest Asia *e.g.* Miller Rosen, 1989; Saeedi, 2010). The premise that micro-finds are categorized as primary refuse makes it easy to interpret the distribution of micro-finds as activity areas. I use this premise to reconstruct cleaning activities by adding the macro-find distribution.

The distribution and density of micro-finds[2] can be compared with that of macro-finds through a rank correlation analysis (Shennan, 2008:pp.145-147; see tab. 1). Here, I separate the amplitude of find densities of macro and micro-finds of a specific type into equal numbers of ranks, so that a comparison of density categories is feasible. I determined the count of ranks per type based on the maximal count of micro-finds per liter, so that bones have more ranks than lithics; the higher a rank, the more ct/l. I assume four ideal relationships between micro- and macro-finds:

1. An area without macro but with micro-finds. In this area activity took place but the remains were cleaned up afterwards.
2. An area with micro as well as macro-finds. This area was used for a specific activity but not thoroughly cleaned.
3. An area without micro but with macro-finds. This is a secondary refuse area and can be considered as one only used for refuse.
4. An area without micro and macro-finds. This area is clean simply because no detectable production or discarding took place here.

2 The border between micro- and macro-finds is somewhat arbitrary and there a lot of different definitions (Rainville, 2005:p.17); in my case the border is at 5 mm.

Ranks of the density in count per litre (ct/l)				
Lithic micro-finds				
Rank 1	Rank 2	Rank 3	Rank 4	Rank 5
0 – 2,75 ct/l	2,76 – 5,50 ct/l	5,51 – 8,25 ct/l	8,26- 11,00 ct/l	11,01 – 13,75 ct/l
Lithic macro-finds				
Rank 1	Rank 2	Rank 3	Rank 4	Rank 5
0 – 0,244 ct/l	0,245 – 0,488 ct/l	0,489 – 0,722 ct/l	0,723- 0,966 ct/l	0,967 – 1,200 ct/l
Bones micro-finds				
Rank 1	Rank 2	Rank 3	Rank 4	Rank 5
0 – 7 ct/l	8 – 14 ct/l	15 – 21 ct/l	22 – 28 ct/l	29 – 35 ct/l
Rank 6	Rank 7	Rank 8	Rank 9	Rank 10
36 – 42 ct/l	43 – 49 ct/l	50 – 56 ct/l	57 – 63 ct/l	64 – 70 ct/l
Bones macro-finds				
Rank 1	Rank 2	Rank 3	Rank 4	Rank 5
0 – 0,14 ct/l	0,15 – 0,28 ct/l	0,29 – 0,32 ct/l	0,33 – 0,46 ct/l	0,47 – 0,60 ct/l
Rank 6	Rank 7	Rank 8	Rank 9	Rank 10
0,61 – 0,74 ct/l	0,75 – 0,88 ct/l	0,89 – 1,02 ct/l	1,03 – 1,16 ct/l	1,17 – 1,30 ct/l

Table 1. The ranks of the micro and macro-finds from bone and lithic debris at Fıstıklı Höyük.

With these considerations in mind, I can reconstruct what objects were removed from what context and, therefore, what was considered unclean. Based on the absence and presence of specific objects, I can approach an emic category and, even though I do not know the symbolism behind it, I can still describe it.

Elias' understanding of (un)cleanness and its connection to self-discipline can roughly be summarized as follows: the more self-discipline exists within a society, the stricter people will construct their emic categories. The less self-discipline is imposed on people, the more open these categories become. Therefore, in relation to evaluating (un)cleanness, the more internalized self-discipline exists within a society, the more rules, categories and evaluations are expected to exist around the topic of cleaning and cleanliness. I can explore these premises by comparing different case studies of cleaning activities. As Elias' work was a long-term study, it helps me to compare how strictly people cleaned over the course of centuries. Unfortunately, with Fıstıklı Höyük, I only have one single example and so Elias' ideas are more intuitive, rather than analytically influential, for the present work. Nevertheless, I can compare the results in Fıstıklı Höyük with my own, in Elias' sense, strongly self-disciplined, Western European society. For future research, however, a more analytical comparison could be valuable.

Cleanliness and Uncleanness in Fıstıklı Höyük

Research on (un)cleanness, cleaning activities and emic categories in archaeology requires investigation at the scale of the settlement. I investigate primary and secondary refuse, and the documentation of micro-finds is especially important to my methodological approach. For this reason, I will present a single, well documented site that suits the criteria for this analysis: the late Neolithic site of Fıstıklı Höyük.

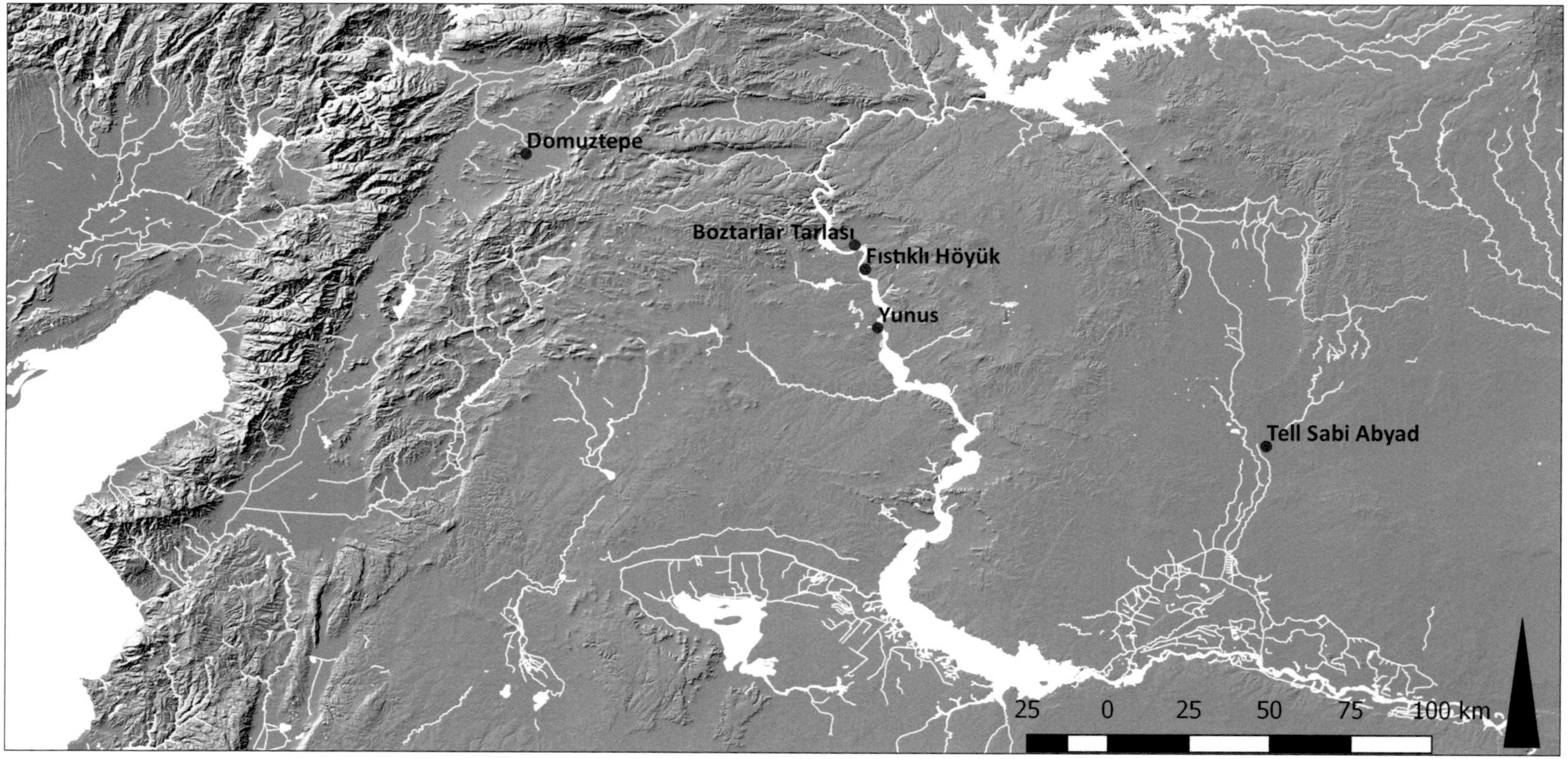

Figure 1. Localization of sites in western upper Mesopotamia mentioned in the text (Drawings by George Cyrus).

Fıstıklı Höyük is a small site along the Euphrates in southeast Turkey (Bernbeck & Pollock, 2003, see fig. 1). It belongs to the so-called Halaf tradition and was occupied briefly from ca. 5980 – 5733 BCE, based on summed calibrated 14C data. The occupation is calculated to have been even shorter (5850 to 5750 BCE) by using the Baysian approach of the Programm OxCal (Campbell, 2007). In general, Halaf ceramics date between ca. 6000 and 5500/5000 BCE (Bernbeck & Nieuwenhuyse, 2013). These ceramics had a large distribution: from the northern Levant and the Taurus Mountains of Anatolia to the Euphrates and Tigris basins of northern Mesopotamia and the Zagros Mountains. Within such a wide area of distribution, settlement patterns seem to vary by region (*e.g.* Campbell, 2003; Campbell, 2004 for Domuztepe in the Kahramanmaraş province, in southeastern Turkey; Akkermans et al, 2006 for Tell Sabi Abyad, ar-Raqqa Governorate, in northern Syria, see fig. 1) and might be an indicator of different forms of organization.

In this article, I will concentrate on the region of the upper Euphrates, where Fıstıklı Höyük lies. Reinhard Bernbeck (2008) developed the concept of multi-sited communities on the basis of the stratigraphy of Fıstıklı Höyük. The occupation pattern at Fıstıklı Höyük changed several times over the course of the 100-250 years of its late Neolithic occupation. At the beginning of Phase IV, a seasonal camp existed; later on in Phase IIIc, the typical small square buildings normally interpreted as storage buildings were added. In the following two phases, IIIb and IIIa, five *tholoi, i.e.* round buildings usually interpreted as living spaces, were erected. Finally, after a short time of occupation, the inhabitants abandoned the site. Sometime afterwards, the site was reused in the phase IIIx, characterized by post-holes suggesting that the site was used as a seasonal camp again.

Fıstıklı Höyük is not the only briefly occupied settlement in the Halaf tradition; the sites of Yunus (Woolley, 1934) and Boztarla Tarlası (Algaze, Breuninger & Knudstad, 1994:pp.36f.) in the surrounding area are also rather small and were probably as equally short-lived. Furthermore, sites like Tell Sabi Abyad are interpreted as composites of multiple small sites (Akkermans et al, 2006). Thus, archaeologists observe small sites with short occupation phases and substantial changes from phase to phase. Hence, Bernbeck (2008) concluded that communities changed their residence frequently and used multiple locations at the same time; these communities were multi-sited.

Maria T. Starzmann, Susan Pollock and Gabriela Castro Gessner conducted investigations concerning the modes of production in Fıstıklı Höyük. Starzmann (2012) reconstructed several communities of practice on the basis of the lithic assemblage and showed a high degree of improvisation and reuse in lithic production. Pollock and Castro Gessner (2009) investigated find distributions in connection to the architecture and installation. They concluded that at least certain aspects of production were performed collectively and in public.

In summary, Fıstıklı Höyük was a small settlement in a society presumably without settlement hierarchy and thus without central places. The community at the site had a collective mode of production and consumed food together in open spaces, but lived in small round houses with attached storage buildings. They often changed their place of living and can thus be defined as multi-sited communities. My analysis concerns the largest extension of Fıstıklı Höyük, phase IIIa. In the following analysis, I separate the mound into five spatial categories (see fig. 2):

1. The northern production area – this is the public outdoor area where much of the production took place; here, two ovens for food were installed, and there are multiple use surfaces.
2. The western refuse area – this is a dumping area at the western margin of the settlement on the slope towards the Euphrates.
3. The spaces in the *tholoi* – this concerns the five round buildings in the center of the mound interpreted as living space. These rooms were roofed and the surfaces prepared.
4. The cell buildings – four of the five round structures had cell buildings next to them that are interpreted as storage buildings and also had prepared floors.
5. The space between the *tholoi* and cell buildings – these are the unprepared surfaces in the settlement.

The site also features an earthwork in phase IIIa which is not part of my statistical approach. It is interpreted as protection against floods (Bernbeck & Pollock, 2003:pp.31f.), but this is contradicted by the position of the earthwork. Especially in the northern production area, the excavators sampled micro-finds, so that I can compare the micro- with the macro-finds. For this paper, I will concentrate on the lithic debris and bones because they brought the most significant results.

By comparing the ranks of the macro and micro-finds I can test the four ideal relationships I established earlier. Lithic density is separated into five ranks while bone density has ten ranks. In the northern production area, the samples originated in the surfaces. The micro-finds consisted mostly of bones (ranks seven to ten) and lithic debris (ranks three and five). This indicates that the production of lithic tools and food took place here. The macro-finds are similar in terms of bones (up to rank eight) but there is less lithic debris (ranks two to three). Thus, I conclude that bigger lithic debris was cleaned away, whereas fragmented bones were not. While bones were not considered unclean, lithic debris definitely was. The samples from the space between the buildings at the center of the mound also originated from unprepared surfaces. There, almost no micro-finds (*e.g.* ranks one to two for lithic) were found, so there is no evidence for lithic production there. However, in the narrow gaps between the buildings a lot of secondary refuse in form of lithic macro debris was found (ranks one to five for lithic but just ranks one to two for bone). So, this area was used as a refuse area for lithic production debris.

The western refuse area resembles the one just described. In this area, no samples for micro-finds were taken, but I assume that this was not a food production area because there are no installations such as those in the northern area. Whether lithic tools were produced in the midden area is unclear, but it was a refuse area at the margin of the settlement.

Inside the buildings there were relatively few micro-finds, but the samples originated from prepared surfaces that are less permeable and therefore easier to clean. The cell buildings were almost completely empty, with nearly no micro (ranks one to two for lithic debris and bone) and only a few macro-finds (ranks one to three for lithic debris and rank one for bone) retrieved therein. This pattern again underlines the interpretation of storage buildings that were not used as a production place or refuse area. The *tholoi*, too, had nearly no macro (ranks one to three for lithic debris and rank one for bone) and micro-finds (ranks one to two for lithic debris and bone), so I have no evidence for production and refuse there. Only burned clay was found as an indicator of fire; in this case, a fire without cooking refuse could be interpreted as a fire for warmth or light.

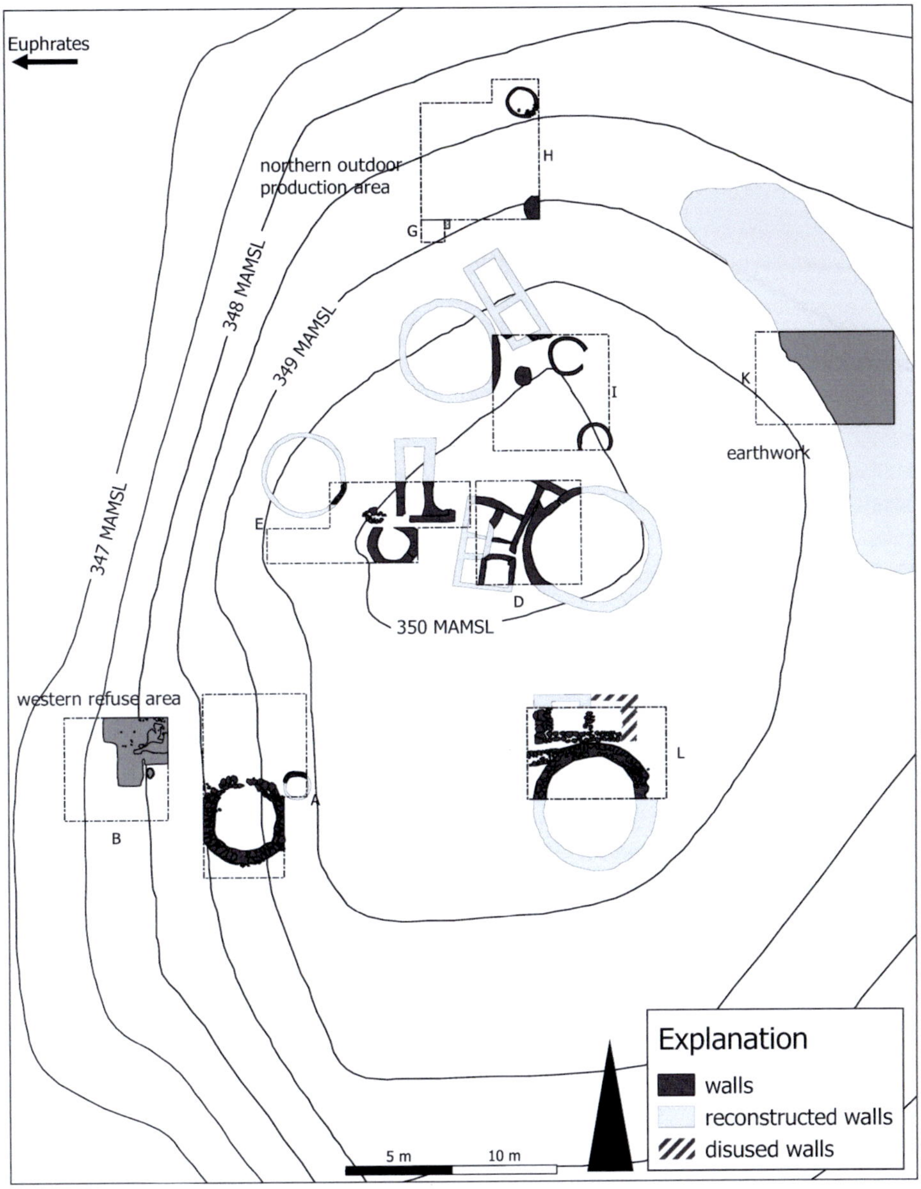

Figure 2. Fıstıklı Höyük in phase IIIa, with reconstruction following Bernbeck et. al. 2003 (Drawings by George Cyrus).

Using Douglas' definition of uncleanness, it is now possible to redefine the etic analytical categories with respect to the reconstructed cleaning activities. In the village of Fıstıklı Höyük, there are four different treatments of space visible in the archaeological record that might mirror those emic categories (see fig. 2). The first is the treatment of the northern production area. There, lithic tools and food were produced, but lithic debris was considered unclean and was accordingly removed. Secondly, there is the refuse area in the narrow gaps between the buildings and on the western slope of the settlement; there, objects were discarded that had been cleaned out of other places. Such an area is not necessarily a deposit of trash but can also be viewed as some type of storage place for recyclable objects - after all, it is located in the settlement and the excavators found reused objects at the site. Thirdly, the cell buildings were clean and used for storage of other goods, probably food. Fourthly, inside the *tholoi*, fires were meant to heat up the space; this might point to the interpretation that they were used as sleeping areas during colder seasons.

None of these patterns seem to point towards social structures that produced a large degree of internalized social pressure in the form of self-discipline. At the northern production area, for example, many macro-finds were scattered, and it seems that the inhabitants did not consider cleaning up this place. Furthermore, most of the clean areas lacked micro-finds except for burnt clay, suggesting that the use of these areas was limited to activities that didn't leave traces in the archaeological record. Furthermore, people did not dedicate much effort to relocating lithic refuse: the dumping areas between the houses and on the western slope lay right next to the identified production area. With Elias in mind, the archaeological finds of the Fıstıklı Höyük settlement could be interpreted as a culture and community which inflicted less self-discipline in its population than modern Western societies today. This claim is further confirmed by the non-centralized settlement patterns and the creative recycling of lithic and pottery debris.

Final Reflections

In this article, I started with the statement that uncleanness is culturally specific and not a natural category of hygiene. I based these claims on the definition of uncleanness as presented by Douglas and subsequently tried to connect it with Schiffer's definition of cleaning. Because uncleanness is not a category, but rather a border-crossing, this research cannot provide a list of objects that are deemed unclean by way of conclusion. What is, and is not unclean is contextual. Archaeological contexts are not complete records but rather fragmented (Lucas, 2012:pp.18-73) and, after nearly 8000 years, most organic materials have decayed. In my case study, I therefore concentrated on bones and lithic debris. The reconstruction of emic categories is framed by the fragmentation of the record. Furthermore I have used descriptive statistics to evaluate the cleaning behavior and my interpretation is limited to the results from Fıstıklı Höyük itself. I can compare the cleaning patterns of lithic and bones on the site, but I have no independent reference. Fıstıklı Höyük could be the norm as well as the exception, however the structural similarity to other halafian sites is an argument for viewing Fıstıklı Höyük as an average site.

I have nevertheless been able to describe some aspects of the emic categories that include the preserved finds. As I have tried to demonstrate, the people of Fıstıklı Höyük carried out productive activities together in the northern area of the settlement. The analysis of the micro-finds, however, adds a new element to this production area. Following the implications of the micro-debris analysis, this was surely a production place for lithics and food; no local separation could be detected. But whereas fragmentary bones were left where they had been processed, the lithic refuse was cleaned away and brought to designated refuse areas. This observation implies that although both activities were carried out at the same location, the refuse was nonetheless treated differently: lithic refuse was considered unclean while bone fragments were not. Here we touch on the emic view on uncleanness, but it remains purely descriptive. The reason why lithics were viewed problematically cannot be reconstructed from the archaeological record. On a functional level, lithics could be interpreted as dangerous, because of their sharp edges, but this is up to speculation.

Finally, this leaves us with the somewhat paradoxical situation that it is possible to describe emic categories of uncleanness in prehistoric societies by resorting to etic perspectives. It still remains vague, however, as to why something was considered unclean. In this regard Elias approach could help to continue research. A comparison between different categories of uncleanness would be interesting: inspired by Elias long-term study we could investigate the degree to which past societies were 'disciplined' and how they changed over time, and we could correlate these changes to the concept of uncleanness with other social changes. Me must not, however, uncritically adopt Elias' conclusion, but rather take his work at a starting point. Today, Elias' concept of the civilizing process forms part of intellectual history and needs to be viewed in the context of its time.

Of course, different sites will require different methods depending on the state of documentation and preservation and my investigation depended strongly on the documentation of micro-debris. Still, I hope to have shown that the reconstruction of emic views of uncleanness is possible and profoundly connected to emic categories, as well as to often overlooked cleaning activities. If we want to reconstruct the lives of people in past societies, engaging with questions of uncleanness is indeed a worthwhile endeavor.

Acknowledgments

Scientific investigations and papers are always collaborations. I first thank Susan Pollock and Eva Rosenstock who helped me to analyze the micro-finds during my master's thesis and always had time for my questions and doubts. I am also thankful for the critiques and tips from Joseph Rebling, Moslem Mishmastnehi, Judith Stern, Giulia Russo, Stefan Schreiber and Arno Hölzer that helped shape my paper from a nearly unreadable draft into the form you are now looking at. Finally, I thank the anonymous reviewer for their constructive critiques.

Bibliography

Akkermans, P. M. M. G., Cappers, R., Cavallo, C., Nieuwenhuyse, O., Nilhamn, B. & Otte, I. N. (2006) Investigating the Early Pottery Neolithic of Northern Syria: New Evidence from Tell Sabi Abyad. *American Journal of Archaeology*, 110, 123 – 156.

Algaze, G., Breuninger, R. & Knudstad, J. (1994) The Tigris-Euphrates archaeological reconnaissance project: Final report of the Birecik and Carchemish Dam survey areas. *Anatolica*, 20, 1- 96.

Bal, M. (2011) Interdisciplinarity: Working with concepts. *Filolog*, 3, 11 – 28.

Bernbeck, R. (1997) *Theorien in der Archäologie*. Tübingen, A. Francke Verlag.

Bernbeck, R. (2008) An archaeology of multisited communities. In: Barnard, H. & Wendrich, W. (eds.) *The archaeology of mobility. Old world and new world nomadism*. Cotsen Advanced Seminar series 4. Los Angeles, Cotsen Institute of Archaeology, University of California, pp. 43 – 77.

Bernbeck, R. & Nieuwenhuyse, O. (2013) Established paradigms, current disputes and emerging themes: the state of research on the late Neolithic in upper Mesopotamia. In: Nieuwenhuyse, O., Bernbeck, R., Akkermans, P. M. M. G. & Rogasch, J. (eds.) *Interpreting the late Neolithic of upper Mesopotamia*. Palma Egyptology Papers on Archaeology of the Leiden Museum of Antiquities 9. Turnhout, Brepols, pp. 17 – 37.

Bernbeck, R. & Pollock, S. (2003) The biography of an early Halaf village: Fıstıklı Höyük 1999 – 2000. mit Beiträgen von Susan Allen, Ana Gabriela Castro Gessner, Sarah Kielt

Costello, Robert Costello, Melissa Foree, Margarita Y. Gleba, Marie Goodwin, Sarah Lepinski, Carolyn Nakamura und Sarah Niebuhr. *Istanbuler Mitteilungen*, 53, 9-77.

Binford, L. R. (1962) Archaeology as anthropology. *American Antiquity*, 28 (2), 217-225.

Campbell, S. (2003) Domuztepe 2003. *Anatolian Archaeology. Reports on research conducted in Turkey*, 9, 4-6.

Campbell, S. (2004) Domuztepe 2004. *Anatolian Archaeology. Reports on research conducted in Turkey*, 10, 4-6.

Campbell, S. (2007) Rethinking Halaf chronologies. *Paléorient*, 33 (1), 103-136.

Douglas, M. (1994) *Purity and danger. An analysis of the concepts of pollution and taboo.* London, Routledge.

Elias, N. (1939) *Über den Prozess der Zivilisation. Soziogenetische und Psychogenetische Untersuchungen. Zweiter Band. Wandlungen der Gesellschaft. Entwurf zu einer Theorie der Zivilisation.* Basel, Verlag Haus zum Falken.

Fladmark, K. R. (1982) Microdebitage Analysis: Initial Considerations. *Journal of Archaeological Science*, 9 (2), 205-220.

Foucault, M. (1977) *Discipline and punish. The birth of the prison.* London, Penguin.

Gifford, D. P. (1978) Ethnoarchaeological observations of natural processes affecting cultural materials. In: Gould, R. A. (ed.) *Explorations in Ethnoarchaeology*. School of American research. Advanced seminar series. Albuquerque, University of New Mexico Press, pp. 77-101.

Gifford, E. W. (1916) Composition of California shellmounds. *University of California Publications in American Archaeology and Ethnology*, 12 (1), 1-29.

Harris, M. (1979) *Cultural materialism: The struggle for a science of culture.* New York, Random House.

Hayden, B. & Cannon, A. (1983) Where the Garbage Goes: Refuse Disposal in the Maya Highlands. *Journal of Anthropological Archaeology*, 2 (2), 117-163.

Lucas, G. (2012) *Understanding the archaeological record.* Cambridge, Cambridge University Press.

Miller Rosen, A. (1989) Ancient town and city sites: A view from the microscope. *American Antiquity*, 54 (3), 564-578.

Moore, H. L. (1982) The interpretation of spatial patterning in settlement residues. In: Hodder, I. (ed.) *Symbolic and Structural Archaeology*. New directions in archaeology. Cambridge, Cambridge University Press, pp. 74-79.

Murray, P. (1980) Discard location: The ethnographic data. *American Antiquity*, 45 (3), 490-502.

Pike, K. L. (1967) *Language in relation to a unified theory of the structure of human behavior*. 2nd ed. The Hague, Mouton & Co.

Pollock, S. & Castro Gessner, A. G. (2009) Engendering communities: The contexts of production and consumption in early Mesopotamian villages. In: Terendy, S., Lyons, N. & Janse-Smekal, M. (eds.) *Que(e)rying archaeology: Proceedings of the 37th Annual Chacmool Conference*. Calgary, Archaeological Association, University of Calgary, pp. 240-249.

Rainville, L. (2005) *Investigating upper Mesopotamian households using micro-archaeological techniques*. British archaeological reports. International Series 1368. Oxford, Archaeopress.

Saeedi, S. (2010) Microdebris analysis. In: Pollock, S., Bernbeck, R. & Abdi, K. (eds.) *The 2003 excavations at Tol-e Baši, Iran. Social life in a neolithic village*. Archäologie in Iran und Turan 10. Mainz, Verlag Philipp von Zabern, pp. 246-255.

Schiffer, M. B. (1972) Archaeological context and systemic context. *American Antiquity*, 37 (2), 156-165.

Schiffer, M. B. (1976) *Behavioral archeology*. Studies in Archeology. New York, Academic Press.

Scott, J (2017) Against the Grain. A Deep History of Earliest States. New Haven, London, Yale University Press.

Shennan, S. (2008) *Quantifying archaeology*. 2nd ed. Edinburgh, Edinburgh University Press.

Starzmann, M. T. (2012) Die Lithikfunde vom Fıstıklı Höyük. Technologie und Handwerk in der Halaf-Zeit. *Zeitschrift für Orient-Archäologie*, 5, 44-73.

Trigger, B. G. (1989) *A history of archaeological thought. Cambridge*, Cambridge University Press.

Wolfe, N. D. & Panosian Dunavan, C & Diamond, J (2007) Origins of Major Human Infectious Diseases. Nature 447, 279 - 283.

Woolley, L. (1934) The prehistoric pottery of Carchemish. *Iraq*, 1 (2), 146-162.

CARPE Dirt, Disease, and Detritus: Roman Sanitation and its Value System

Ann Olga Koloski-Ostrow

I. Introduction

This article does not seek to undermine the accomplishments of the Romans in any way. Their magnificent architectural feats and their impressive technological accomplishments in engineering and urban infrastructure, including water supply and sewer systems in their cities for vast numbers of people, stand on their own merits. We do, however, seek to go beyond the facts of Rome's beautiful monuments and greatness to probe the dark and dirty reaches of Roman antiquity and to seek truths about the realities of Roman daily life.

In order to grasp the daily realities of the Romans with more clarity, we must use whatever evidence we can find, without viewing things through the lens of present-day values, whether we are considering politics, religion, societal structures, economics, gender and sexuality, or more subtle concepts, such as what constituted health and cleanliness in the Roman world. The key to success – in this case revealing the realities of Roman sanitation – will always involve treating the Romans on their own terms. This strategy may be quite difficult, given the evidence we have and the unexpected differences we may find between contemporary Western daily life and that of the ancient Romans.

We might say, in fact, that the archaeological evidence of elaborate urban structures and military strength, as well as refined systems of governance and worship, risks representing the Romans as very much similar to a modern society – especially for researchers working in the Western academic tradition. Because the Romans developed such sophisticated methods for managing their waste (drains, sewers, open channels, well-made pipes, and various kinds of highly functional toilets), we could end up assuming a gloss of cleanliness that did not really exist but is rather imagined based on the sanitary amenities in the archaeological record. The Romans also put tremendous effort into the provision of fresh water for their cities by way of elaborate aqueduct systems. They appear to have been masters at constructing urban infrastructures for removing dirty water from domestic and public facilities such as walkways, roads, and enormous bath complexes, and for laying out both public and private latrines and toilets across the urban landscape. For centuries, archaeologists have interpreted these structures as proof of a heightened cultural awareness and estimation

in: Bredenbröker, I., Hanzen, C. & Kotzur, F (eds) 2020: *Cleaning and Value. Interdisciplinary Investigations*, Sidestone Press (Leiden), pp. 105-126.

for cleaning and cleanliness – in modern terms, sanitary practices and sanitation. I hope to remove some misunderstandings derived from these assumptions and to consider some new ways to think about so-called Roman "sanitary" structures. Once again, this exploration is done not to denigrate the accomplishments of the Romans, but in order to achieve a better understanding of their daily life that comes closer to how cleanliness and sanitation were used and valued in ancient Rome.

I have written on a wide range of Roman urban problems related to smell in architectural infrastructure, housing, small-scale industries, entertainments, religious buildings and rituals, and in the disposal of the dead (cremation, in particular) (Koloski-Ostrow, 2015a). In this paper, I would like to focus on some of these urban problems specifically in relation to urban sanitation. I hope to challenge common modern perceptions of Roman social customs, to probe Roman beliefs about health, to understand more about the Roman tolerance for filth in their cities, and to investigate various Roman attitudes, insofar as we are able to uncover them. In this process we shall see how easy it is to misinterpret matters surrounding Roman hygiene and sanitation from a contemporary Western point of view, given that urban structures in the Western world today seem to share somewhat similar systems to Roman sewers and water supply mechanisms. My evidence draws heavily on the archaeological remains of architecture, urban layout, streets, and graffiti in Pompeii, Herculaneum, Ostia, and Rome and on a variety of ancient literary texts. With this evidence, I explore ancient daily life in relation to sanitation and its value to the Romans.

II. Sewers and Sanitation

In the contemporary Western world, sewer systems are tightly tied to urban sanitation. Today, we build sewers for the purpose of removing human waste and improving the cleanliness of cities. Because Roman and contemporary Western sewer systems appear to be quite similar, it is easy to fall into the assumption that the Romans built sewers for the same reasons, i.e., for removal of urine, excrement, and rotting garbage. Closer examination of those parts of the Cloaca Maxima (Reimers, 1991) – the great sewer of Rome – that are still accessible clearly shows, however, that not much thought was given in its construction to what we would call sanitary engineering.

In fact, the Cloaca Maxima primarily drained and removed dirty water that would pool up on the uneven streets of Rome and in low-lying areas, especially after flooding of the Tiber River which would happen with striking frequency (Dio. 39.61.1-2)[1]. The insight that drainage was the primary function of the Roman sewer (which seems to have been true for sewers in other Roman cities as well (Poehler, 2012)[2]) helps to paint a much more likely picture of the "Roman" perspective on issues around health and sanitation.

While the ancient city of Rome did have wells and springs, the Tiber River served as Rome's main water supply. Despite the Tiber serving as the city's most important fresh water supply – at least before the aqueducts were introduced – it appears as if no one paid much

---

1 Cassius Dio (39.61) describes a flood in Rome in the year 54 BCE. For other inundations of the Tiber in Roman times, see Lugli, 1934:pp.231ff. and also Aldrete, 2006: pp. 91-165. Aldrete describes both the immediate and delayed effect of floods on living conditions and sanitation.

2 The article focuses on the mechanisms for water removal at Pompeii. Pools of standing water are well attested in our sources for seventeenth and eighteenth century Paris. Cf. Corbin, 1986:pp.31-34. For detailed references to the sewers of Paris in the Old Regime, see Lemoine, 1929-1930 and Reid, 1991:p.187 and especially ns. 14 and 15.

attention to the polluted detritus being dumped into the river. We do have some literary evidence to verify the problems Romans could face when an 'open sewer', namely a river, was running through the middle of a town. Pliny the Younger writes to the Emperor Trajan concerning an open sewer in Amastris (in Turkey), which was essentially a river completely clogged with sewage and other debris: "*ita pestilens est odore taeterrimo*", "such pestilence is (accompanied) by the most disgusting stench". When Pliny offers to cover this open sewer he says nothing about improving sanitation, but instead comments on the need to cover over this repugnant eye-sore which also emitted a horrific stench (Pliny, *Ep.* 10.98, 99).

So, what does this exchange tell us about Roman sanitary practice in relation to the water supply? If sanitation involves reducing the risk of the spread of pests and disease as well as the management of powerful smells, Roman sewers did not measure up to the challenge of staving off the problem. With regard to the great sewer of Rome, we can say that it moved odoriferous water away from where it hindered cleanliness, economic growth, urban development, and even industry, but the sewer itself did not contribute much at all to improved standards of urban sanitation (Koloski-Ostrow, 2015b:pp.38-51).[3] My work in the sewers of Herculaneum and Pompeii have brought me to the same conclusion (Scobie, 1986:pp.408; Koloski-Ostrow, 2015, Fig. 92; Jansen, 2011; Camardo, 2011).

Discoveries of repulsive deposits of hardened muck and sludge, such as the one under Cardo V at Herculaneum (Maiuri, 1958:p.469) and in various sewers around the Roman world strongly support the assumption that urban sewer systems provided minimal overall sanitary benefits, at least as we presently understand them. Several ancient sources tell us that Roman cloacae needed manual cleaning from time to time, as they do today. Pliny's *Ep.* 10, for example, mentions prisoners conscripted to clean the sewers (Pliny, Ep. 10.32).[4] The Cardo V deposit in Herculaneum measures about 1.35 meters high and no amount of water, however fast it flowed, was ever able to remove it. Apparently ancient sewer cleaners found it too difficult to remove as well. As a result, the blockage remained in the sewer channel until excavators found it in the 1920s.

The sewers of Pompeii have not yet been fully explored, but two sewer lines apparently serviced two sets of the city's public baths (Forum and Stabian), and their main purpose was to remove substantial quantities of dirty water, which in turn flushed down some public toilets and out of the city. Many other parts of Pompeii simply had no sewers so filth, manure, and dirty water were simply washed down the sloping streets by rain at whatever pace it fell (Poehler, 2012).[5]

---

3 You find here maps, plans, and extensive discussion of the layout of the Cloaca Maxima in Rome. See also Hopkins, 2012:pp.81-102. For older sources on the Cloaca: Bauer, 1989:pp.43-67; Bauer, 1993:pp.288-290; Reimers, 1989:pp.137-141; and Reimers, 1991.

4 Pliny mentions the cleaning of sewers by prisoners conscripted for this odious job. Ulpian's Dig. 43.23. 1-2 records an edict of a praetor in Rome stating that the sewers were to be kept clean and in a good state of repair because they might cause the collapse of buildings (ruinas minantur immunditiae cloacarum). We can assume that major clogs in the Cloaca might have led to a backup of water, which eventually would scour out the foundations of cheaply constructed buildings, therefore causing collapses. Not surprisingly, we can find examples of sewer clogs in our modern world as well. The most impressive of these, perhaps, is the great London grease clog, but examples have been found in the sewers of Belfast, Denver, and Melbourne as well: www.theguardian.com/culture/2018/feb/04/fatberg-museum-london-display-pickling-age-waste

5 I have come to these conclusions via personal observation. The overall drainage system, however, is more sophisticated and complex than these observations make clear.

The sewers of Ostia are also generally unexplored, but we have at least been able to identify many of the sewer covers which allow us to trace most of the main sewer lines. The town was very close to the Tiber River so flooding was a constant problem (Jansen, 1995; See also: Jansen, 2002).[6] Given these conditions, construction of cesspit toilets at Ostia was hardly practical, although archaeologists do occasionally find cesspits in Ostian apartment buildings. The Ostian sewer system very likely received more excreta directly from the streets mixed with water runoff on the streets, than directly from house or apartment toilets. In fact, Roman toilets were, as it turns out, not always, and not necessarily, connected to city sewers. The lack of sewer connections, when we identify them, proves a new and different purpose for many Roman sewers, which were not tied directly to urban sanitation as we understand it.

III. Toilets and Sewers Connections

Archaeological excavations have identified a clear preference for internal cesspit toilets in the domestic dwellings of Pompeii (fig. 1) and Herculaneum (fig. 2) in the first century CE.

These toilets – one-seater arrangements, for the most part – had no sewer connections (Koloski-Ostrow, 2015b, Figs. 1, 2, 28, and 35). While working on my book on the archaeology of sanitation in Roman Italy I had plenty of occasions to test these theories by searching for and not finding sewer connections.

Ulpian's *Digest*, written between 211 and 222 CE, affirms that sewer connections from private dwellings were legal, so one must wonder why so few property owners chose to have their facilities connected to the public *cloacae*? (Ulpian, *Digest*, 43.23.1-2.) Private house toilets in Pompeii and Herculaneum were almost always located right in or near the kitchen area where food was prepared (fig. 3).

The soothing smells of a hearty stew would have intertwined with the sickly odours from the nearby open cesspit in the kitchen. We can surmise that the contents of the cesspits were saved in order to sell to farmers for agricultural fertilizer or to be used as fertilizer right in household gardens (a direct, ready source for the family). Below, we explore some other explanations for this situation (Varro, *RR* 1. 13 4.; Columella, *De Re Rustica* 1.6.24).

Since Roman sewer openings had no traps, one could never tell what might climb out of an open sewer and into the pipes leading up into a house. The Romans were apparently leery about what this might imply and one dramatic ancient story illustrates the danger of owning a house with a connection to a public sewer in the first or second centuries CE.

The author Aelian (a writer of the second century – 175-235 CE) tells us about a wealthy Iberian merchant in the city of Dichearchia (Roman Puteoli, modern Pozzuoli) who discovers that every night a giant octopus would swim into the sewer from the sea, up through the house drain pipes right into and then out of the toilet in order to eat all the pickled fish stored in his well-stocked pantry (*De Natura Animalium*, 13.6). The house slaves bludgeoned the hungry octopus to death, but not before it devoured a significant amount of pickled fish and terrorised the whole household. Would anyone want his house toilet connected to the sewer with such a possibility looming down the drain? Even though Aelian's story is hyperbolic and likely exaggerated for literary effect, urban myths (giant

6 I have worked with Jansen at all three cities, Pompeii, Herculaneum, and Ostia. She crafted maps of sewers in all three cities, which she has allowed me to use in my work. See Koloski-Ostrow, 2015b, Figs. 91, 92, 93.

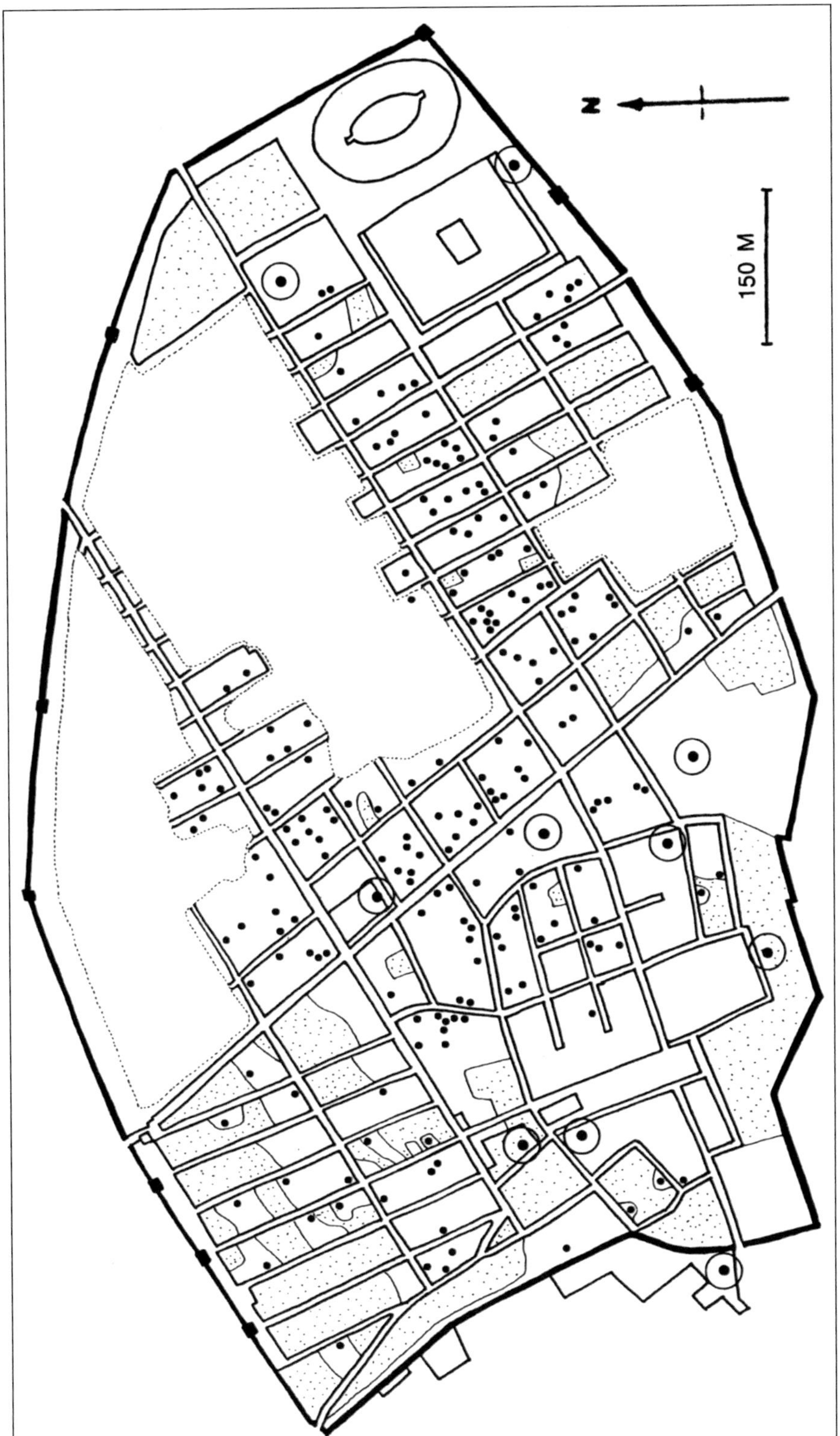

Figure 1. Map of Pompeii with public and private toilets indicated (Photo: Jansen, with permission).

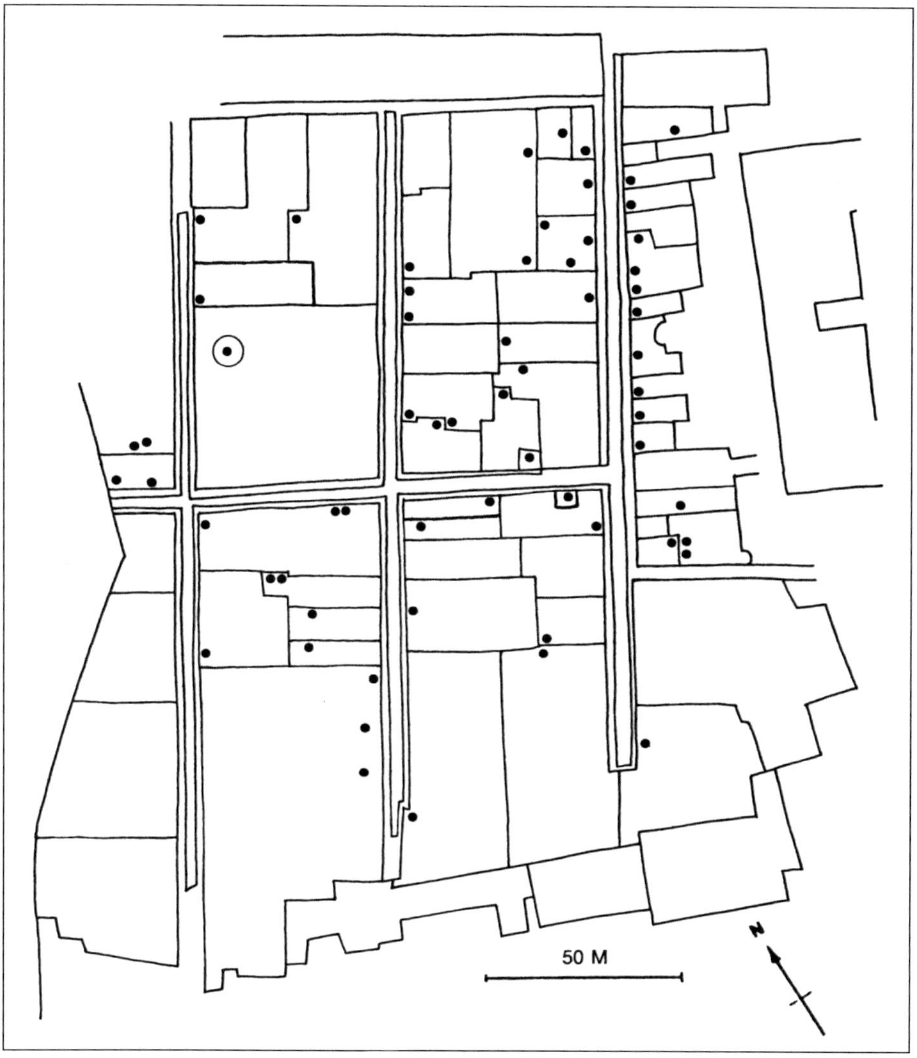

Figure 2. Map of Herculaneum with public and private toilets indicated (Photo: Jansen, with permission).

alligators in the sewers of New York City) still persist today and in some situations have actually been verified. Every once in a while, especially in New York, we read about baby alligators, rats, or certain resilient types of snakes that live and thrive in the sewers, occasionally finding their way into the toilet bowls of apartments (Gee, 2016; Fishbein, 2016). Even with sewer traps in place, these animals can sometimes surprise unsuspecting New Yorkers. The situation in ancient Rome would have inevitably been even more scary.

Roman public latrines, multi-seater toilets that were almost always connected to the main sewer lines of a city, also posed serious threats to users. They were notorious for terrifying customers with flames exploding from their seat openings. Gas explosions of hydrogen sulphide (H2S) and methane (CH4) still destroy sewer covers in cities today (Knowlton, 1932; Popielarz 2015). Furthermore, clients in latrines would have had to worry about the rats and other small vermin threatening to bite their bottoms (Gee, 2016), as well as the imagined

Figure 3. Reconstruction of a Pompeian kitchen (Photo: Koloski-Ostrow).

threat of demons in those black holes. One late Roman writer tells a particularly exciting story about such a demon. A certain Dexianos (*Life of Saint Thecla, Miracle 7*)[7] was sitting on the privy in the middle of the night when a demon raised itself in front of him with savage and enraged ferocity. As soon as Dexianos saw the 'hellish and insane' demon, he 'became stunned, seized with fear and trembling, and covered with sweat'.

7 See also: Jansen, 2011: pp. 165-166 and Dagron, 1918.

Superstition regarding demons (and animal intruders) would provide another good reason for avoiding connections to the sewers. So, to thousands of uneducated Romans, regular disturbances from gasses in the sewers' (Lanciani, 1890:p.56) in the streets emanating from sewer covers or from the holes in public toilets situated over sewer lines would have contributed to a general urban superstition about the dangers of the sewers and the need to avoid them and any systems connected to them.

IV. Housing and Sanitation

Close observation of the archaeological remains in Roman cities, combined with legal texts and various literary references, can lead us to speculate with some authority about the quality of Roman housing in relation to sanitation (Scobie, 1986). Crowded and impoverished urban habitation in apartment houses (cf. Ostia and Rome) provided a fertile ground for unsanitary living conditions (Scobie, 1986: p.427). Sloppily constructed apartment buildings would have had heavy growths of mould and other furry fungi, excessive dampness, residual dust and smoke from charcoal fires, wells open to all apartment dwellers that were often stagnant and polluted, and clogged and overflowing cesspits contributing to the spread of health hazards. Perhaps my visualization will be challenged as pure speculation, but below I confirm with some archaeological examples that such living conditions did exist in Roman houses. While not common for all houses, the occasional house in Pompeii (V. 2. 1, House of the Silver Wedding, for example) would have a drain leading directly from the house toilet (a kitchen toilet, in this case) into a garden, located just behind it to the west. Even though fairly rare, we can imagine that this arrangement would have created a persistent stench in the gardens of such houses.[8]

Tenement buildings, such as those located in Ostia and also in the partial ruin of a tenement surviving at the base of the Capitoline Hill in Rome – the so-called Casa di via Giulio Romano (Packer, 1972) – provide enough archaeological evidence to reconstruct some of the grimy circumstances mentioned above.

Since the ground floors of these multi-story buildings were ringed with countless small shops or manufacturing outfits occupying small spaces on their exterior walls, the smells of fire and smoke from burning ovens and odours from both food processing and food deterioration must also have suffused the rooms of the apartments on the upper floors. Such conditions are still observable in the archaeological record at both ancient Pompeii and Ostia.

Most of these conditions in *insulae* would have also been true for cheap lodgings and inns. While the evidence on-the-ground for such flimsy institutions is modest, one vivid graffito (found in VIII 6, 1.9, a bakery at Pompeii) hints at limited services and resulting discomforts: '*Miximus in lecto, fateor peccavimus, hospes, Si dices quare, nulla metaella fuit.*' (De Vos & De Vos, 1982:p.59).[9] The graffito suggests that, if no modest toilet facilities were provided in a guesthouse, clients could expect to find at the very least a chamber pot in their rooms.

---

8 The issue of septic or sewer odours is still very relevant today. Factors such as humidity, construction, and weather conditions affect mould spores and gas odours. If the cess removed from the kitchen toilet of the House of the Silver Wedding were "dry" when removed, stench would certainly have been diminished. Since it was a kitchen toilet, however, discarded slop must have been poured into the cess regularly, keeping the contents quite moist. Based on some assumptions, then, upon emptying the cess stench could not have been pleasant, even if it did not last that long. See InspectAPedia (2011) for further discussion.

9 'We have wet the bed, I confess, guest; if you ask why: there was no chamber pot (in the room)' The translation is my own.

Figure 4. Pompeii, terracotta downpipe (Photo: Koloski-Ostrow).

Even in the homes of the wealthier Romans – that is, in free-standing, private houses – leakage from the downpipes of toilets located on upper floors would have posed another kind of unsanitary danger (fig. 4).

The fittings of terracotta downpipes loosened over time and their contents would have spilled down the outside of the pipes causing stink and the potential for the spread of e-Coli and other diseases born from contact with excrement. At least fifteen upper storey toilets have been identified at Pompeii and others at Herculaneum and elsewhere (Hobson, 2009:pp.71-77).

V. Industries and Sanitation

In Roman businesses and production contexts, we can note various instances in which excrement and urine mingled very closely with products and procedures. Today these would be deemed unhealthy and unsanitary.

In the Roman city, the production of bread was a thriving business. Producing flour and kneading and shaping the loaves were activities at the beating heart of the city. Keeping the bakeries clean, however, would have been almost impossible, and the sanitary procedures to follow would have been very limited or completely absent. The archaeological evidence already tells us quite a bit about the sanitary conditions of bakeries (in Pompeii and Herculaneum, in particular). The proximity of animal stalls to the baking process is easily deducible. In addition, we have a literary account from the second century writer, Apuleius, explaining the details of life in a bakery (Apuleius, *Metamorphoses*, 9.12). He describes a small flour mill with a bakery inside it, located in an unidentified Roman city in northern Greece. He tells us about the starving, overworked slaves who toiled there (clearly with no bathroom breaks), the puss-filled welts on their bodies from beatings, their sore, inflamed, and teary eyes, and their filthy and tattered rags for clothes. All of these details bring to life a vivid picture of the smelly and unclean environment of the bakery.

While a description of the bread itself would have been the one bright spot in this text on the drudgery and dirt that was a baker's shop, the archaeological evidence even more clearly enhances our understanding of the unsanitary activities that took place there. Donkeys (fig. 5) defecating all over the floor would have circled the mills to grind the flower, probably until they dropped from exhaustion.

Smoke, fire, dust, and grit (scrapped out of the domed brick ovens just before the bread loaves were inserted), would have filled these spaces. The keeping of horses, mules, donkeys, dogs, chickens, ducks, geese, other fowl, and even dormice somewhere in the main part of a Roman house or shop meant that animal dung and diseases would have spread across the property posing a health hazard for all workers and customers.

Fulleries were workshops for washing and stretching wool clothing and for laundering other fabrics as well. They were big business in Roman cities. Like bakeries and gristmills, these establishments could be located anywhere in the city. Ample evidence from Pompeii allows us to read their physical layout quite well. Clothes were soaked in small tubs, surrounded by low walls, filled with water and a mixture of alkaline chemicals, including urine (Martial 6.93.1.; Suetonius, Vespasian 23. 3).[10] The cloth was trampled, then scrubbed by hand, and finally the liquids were squeezed out. Larger basins in the fullery were then used as rinsing tubs to remove the dirt that had resolved in the first process (Flohr, 2006; Wilson, 2003.). We now know that clothes were sometimes also treated with sulphur in the final process, another powerful chemical substance painful to eyes and noses, and certainly not healthy (Flohr, 2003; Bradley, 2002; Flohr, 2007; Wilson & Flohr, 2011; Flohr, 2013:pp.181-241).

Just like houses, many of these small urban business operations had cesspit toilets in locations such as the kitchen or related service areas. In I 6, 7, the *fullonica* of Stephanus, for

---

10 Includes a reference to a tax on urine, instituted by Vespasian, indicating the value of this smelly commodity within urban life. Meiggs, 1960:p.143, also mentions terracotta jars in the streets for the collection of urine.

Figure 5. Reconstruction of a Pompeian Bakery (Photo: Koloski-Ostrow).

example, the (two-seater) toilet is located directly next to the kitchen and only a few steps away from the main working area (Pesando & Guidobaldi, 2006:pp.103-104). In the House of the Oven (VI 3, 27-28), the toilet is in the kitchen and only accessible from the workshop and the first floor. The location of toilets could vary. In house VII 1, 36-37 containing a bakery but no kitchen, a one-seater toilet was situated directly next to the mill room.[11]

Many other urban industries (such as leather tanning, working wool, looming various fabrics, dyeing textiles, manufacturing pottery and glass, fermenting *garum*, processing Roman fish sauce, or producing wines and oils) were also noisy, smelly, and unsanitary, but baking was carried out in close proximity to animal and human excrement and fulling utilized urine, hence making these workspaces potential health hazards for anyone in the area and in contact with the business' products (Pesando & Guidobaldi, 2006:pp.241-242).

We should note that none of these industries were zoned to any particular area of the city, and we find a similar lack of zoning by social status or by the separation of residential and commercial activities. By keeping poor and underprivileged people living and working beside Roman elites all across the urban landscape, the magistrates of Roman cities encouraged interaction and involvement between and among citizens of different social status (Mignone, 2016). This proved a good plan for promoting cooperation between rich and poor during fires, foods, and attacks. When plagues struck, however, as they frequently did, the plan quickly fell apart, and diseases were

11 For houses that had operating bakeries inside them, see the house with a bakery at VII 1, 36-37, in Pesando & Guidobaldi, 2006, plan of region VII, insula 1, on p. 212, and the House of the Oven, VI 3, 27-28, plan of region VI, Insula 3, on p. 172 with discussion on p. 171.

able to spread rapidly. Elites did not escape death any more than the poor. As such, everyone must have inevitably suffered to a similar degree in regards to health and poor sanitation. Laurence has argued that 'deviant zones', including brothels, bars, cheap inns, and similar outposts for lowlife at Pompeii must have been grouped together for the purposes of keeping them out of sight from well-to-do Romans (Laurence, 1994: p. 87). McGinn rejects Laurence's idea of a 'moral geography' at Pompeii for brothels, and posits instead that brothels were cited for profit (2002). Ellis, more recently, has demonstrated convincingly that all sorts of facilities designed for the production, display, transaction, and consumption of food and other goods and services were also sited for profit above all else rather than out of consideration for morals or sanitation (2004). Roman streets were noisy and bustling, not unlike the inner streets of Naples today – unclean, chaotic, but full of life. By mixing the houses of the rich and poor throughout the city and placing dirty, smelly, and noisy commercial enterprises alongside various types of more wealthy residential architecture, magistrates, whether they were aware of it or not, slowly established long-lasting urban security systems and increased profits for all the shopkeepers. They were not, however, promoting new and better sanitation.

VI. Entertainments and Sanitation

We could cover a wide array of public entertainments and activities set in public architecture (in *fora*, markets, temples, theatres, administrative offices, and other public buildings) under this heading. I will limit myself, however, to two examples – those of baths and amphitheaters – for they are the indicative examples in relation to imagining the state of Roman urban sanitation.

Bath buildings, sometimes cared for by public slaves, would have been filled with odours, scents, colourful paintings and mosaics, all manner of noises and behaviour implying the very opposite of a concern for bodily care and cleanliness. The gleaming Roman baths that tourists visit today in sites like Pompeii and Herculaneum tend to give us an impression of cleanliness and purity that was also upheld by the Romans. When we look more closely at the archaeological evidence for baths and combine it with focused medical texts, especially from the first century CE, a different picture emerges. The baths were indeed remarkable for their technology, their water supply and drainage systems as well as various amenities for heating and cooling the body. Their existence in such numbers across the empire and their impressive array of services for cleaning are not to be dismissed. The dangers that these baths posed for their users, however, must also be considered.

A person sitting in the bathtub of a hot room (*caldarium*) in a small city bath at Pompeii – such tubs could hold between eight and twelve bathers at one time – might have been sitting next to someone suffering from open wounds, lesions, lice, gangrene, worms, diarrhea, gonorrhea, tuberculosis, or worse. The first century author, Celsus, lists such health problems (Celsus, *De Medicina*, I. 3-6). Furthermore, the smell of the bath water in the tub would have been overwhelming by the end of a hot summer's afternoon in the Bay of Naples. The aforementioned list of health problems that bath guests might have suffered from or been confronted with comes from only one author, Celsus. Nevertheless, bathing is featured in other Roman writings on medical care, so these suppositions regarding life in baths are not wildly out of place.

The wall painting from the *frigidarium* of the Sarno Baths (fig. 6) offers a warning about what not to do in the bath; namely not to defecate in the bath waters.

Figure 6. Pompeii, Sarno Baths, painting of defecating pygmy (Photo: Koloski-Ostrow).

This painted advice seems to imply that such behaviour could be expected to occur and that it was necessary to instruct guests of the bath not to defecate in the water. Based on such sources, one can imagine that from inside public baths and toilets the strong smells of human excrement and urine never abated, and they were places where guests could possibly contract an array of infections.

Turning now to Roman blood sports – with apologies that we will only consider the admittedly special case of the Colosseum – we can safely assume that by the time the amphitheatre was in operation in 80 CE, drains from the surface of the arena carried body parts and rivers of blood down into the underground channels. As a consequence, the smells from days of games must have been ghastly – a combination of blood, guts, dead men and animals. In the 100 days of spectacles that accompanied the opening of the Colosseum, we are told that 9,000 animals were killed (Shelton, 1998:p.351). Despite not being a permanent occurrence (although 100 days out of 365 is a significant number), the smell of dead flesh and blood must have still emanated to many nearby neighbourhoods and one may assume that it had a lasting influence on these areas. The mechanisms for moving the animals up to the surface of the arena for the slaughter can been seen very well in the Flavian amphitheater of Pozzuoli (Roman Puteoli) (Bomgardner, 2002; Welch, 2009; De Caro & Greco, 1981:pp.37-53). Arrangements in the amphitheater paid no heed to the health or sanitary risks that the games performed there may have posed to the onlookers, such as animal bites or contamination from blood tainted with rabies. Moreover, these entertainment centres always had inadequate or very scarce toilet facilities for the viewers.

In the realm of Roman temples and shrines, a similar scene of uncleanliness and stench must have also been a major distraction for neighbours in those parts of the city. Religious rituals were held outside in the heat of the day and, like gladiatorial or animal games in the amphitheater, involved significant quantities of blood from animal sacrifices along with the strong smell of rotting meat and blood (Clements, 2015). Every festival day – dozens of which were included in the annual cycle of the Roman calendar – required a blood sacrifice of at least one, and sometimes hundreds of bulls or other animals before a crowd (Rüpke, 2011: Part III: pp. 203-272). We can assume that the smoke and smell of fat burning on open altars would have filled the area and nearby streets on a regular basis. We can also speculate that in hot weather, parasites or other intestinal blights were transmitted when the barbecued meat passed around the crowd was undercooked. Perhaps the prayers to the gods to protect the city or the Roman government may well have included prayers to avoid stomach distress from attending the sacrifice.

VII. Death, Dumps and Sanitation

There are still have many questions about death and burial rituals in ancient Rome that we cannot answer to a satisfactory degree, including the number of daily deaths in Rome or the number of burials that were organized privately or depended upon the services of the state. We do know, however, that burial customs changed in different Roman cities over time from the first century BCE to the end of the first century CE (from cremation to inhumation burials) and that funerary practices were also quite mixed.

Our best evidence concerning death and the funeral business and their relation to sanitation comes from the city of Rome, but the cities of Vesuvius and Ostia provide some useful insights too. We can estimate that between 100 BCE and 200 CE, the cemeteries of Rome had to accommodate nearly nine million burials of one sort or another, whether cremations or inhumations (Scobie, 1986). In terms of the danger of exposure to contagious disease, the impact of these deaths on the city would have been significant. The business of dealing with the dead would have been no small hindrance to public sanitation.

Until ca. 100 CE, that is, for over two hundred years, the Romans in Rome primarily preferred to cremate their dead. Elites burned the dead at their own tombs (at private *ustrinae, crematoria*) and the dispossessed were cremated at large public *crematoria*, after which their ashes and bones were buried in pits. Burning flesh and decomposing corpses by the hundreds and thousands, especially during plague periods, would have had a serious effect on air quality, especially just outside the main gates of Rome. We can postulate that this must have been an ongoing state for most Roman cities during the year, although some scholars have claimed that the smell would not have been lingering (Morley, 2015:pp.113-115).

Roman customs related to the preparation of the deceased for burial required a lying-in-state period for up to seven days but bodies would have been in a considerable state of decay by the time they were removed for final rites (Toynbee, 1971:p.45). Perfumes and incense were in high demand at these elite funerals for good reason (Polybius, VI. 53). The funeral business was lucrative in ancient Rome. It thus deserves some consideration within the context of cleaning and value. Death and dying combine concerns for health, sanitation, waste disposal, and human dignity into one major (urban) problem.

If we adopt a conservative estimate for the urban population of Rome around the time of Augustus – namely 750,000 inhabitants – and postulate an annual mortality rate of roughly forty per thousand, some 30,000 residents died each year, more than eighty people per day on average (Bodel, 2000:pp.128-129). In times of plague, these numbers would have risen dramatically. We know that people of lower status were burnt or buried on the day following their deaths, carried off to public *ustrinae* (places for the cremation of the dead)

(Toynbee, 1971:p.45) or dumped in open pits without ceremony (J. Bodel, 2000:pp.131-133; Hope, 2009:pp.65-96). Such pits would have been a serious threat to public health. Stray animals in the city could enter these pits and eat from, dragging around bones and body parts. We learn, for example, from Suetonius that while Emperor Vespasian was having lunch one day in the Palatine Palace, a dog wandered into the dining room and dropped a human hand at his feet (Suetonius, Vespasian, 5.5.; Scobie, 1986:pp.418-419).

The main evidence for these open burial grounds comes from the Esquiline Hill in Rome. Construction of a new residential area in the nineteenth century laid bare some seventy-five ancient mass burial pits, although new work on this area of Rome casts serious doubt about the extent and meaning of Esquiline pits (Lanciani, 1890:pp.64-65).[12] They were immediately associated with the public potter's field mentioned by Horace in his Satire 1 (Horace, *Satire* 1.8.8-16), and they share many similarities to mass graves for the poor in France and England from the sixteenth to the eighteenth centuries (Bodel, 2000:p.130). For whatever reason, the practice of dumping bodies on the Esquiline was eventually abandoned, presumably late in the first century BCE, perhaps around the time of Augustus, perhaps because of their noxious smells and the superstitious taboos and disgust they caused. In any event, Roman city officials would slowly have come to realise the health risks of keeping the pits open, at least on a superficial level.

Lastly, garbage dumps in ancient Rome developed for different purposes: some for household garbage, for fecal waste, for abandoned bodies, and for commercial rubbish. While the disorganized quagmires (Bodel, 2000) of dead flesh and rotting carcasses known to us from such places as Montfaucon in eighteenth century Paris (Reid, 1991:p.11) existed in Roman cities as well, some ancient dumps were quite organized. Monte Testaccio in Rome, for example, the so-called amphora mountain, consists almost entirely of olive-oil *amphorae* from Baetica, Spain, none produced earlier than 144 CE (Claridge, 2010:pp.402-403). The construction of the hill was not haphazard, but planned with level terraces and retaining walls (also built of amphora shards). The discarded *amphorae* were carried up the hill whole and then smashed on the spot. After disposal, everything was liberally sprinkled with powdered lime, clearly in an attempt to neutralize the stench of the rancid oil (Aguilera Martin, 2002; Blásquez Martinez & Remesal Rodriguez, 1999; Blásquez Martinez & Remesal Rodriguez, 2001; Blásquez Martinez & Remesal Rodriguez, 2003).

Ulpian's *Digest* warns about contaminating public water supplies or covering anyone with dung or mud (Ulpian, (Papinian) 43. 11.1.1.), which again suggests that such behaviour occurred often enough for officials to take notice. At Herculaneum in VI, 11-13, the House of the Black Saloon, on a half column of the peristyle, one inscription records the payment of eleven asses for the removal of ordure: '*Exemta ste(r)cora a(ssibus) XI*' (Scobie, 1986:p.414, n. 117), and another painted notice on the side of the water-tower at the northern end of Cardo IV, just as it enters the Decumanus Maximus, bans the dumping of rubbish in the vicinity of the water-tower (Wallace-Hadrill, 2011:p.291). Very likely, people living in houses or tenement buildings dumped debris wherever they could find a

12 Now see Emmerson, forthcoming, chapter 2. I am very grateful to Professor Emmerson for sharing with me her unpublished manuscript with its close analysis of the Esquiline pits. She has shown that Lanciani's evaluation of the pits is very likely a gross exaggeration.

space on a side street, alley, or in an abandoned lot, and unfortunately, no system was in place for its removal, and no easy way to police the practice existed.

The same movement towards an ordered removal of filth by the sewers seems to have inspired the development of massive urban cemeteries and dumps over time. The separation of the living from the dead was an important feature in cities of the Roman world. That is not to say that all dead Romans immediately found their way into well-ordered cemeteries (Bodel, 2000). Nevertheless, most cemeteries in first-century BCE and CE Rome were outside the city walls. Sewers and cemeteries of eighteenth-century Paris were closely linked with the job of disposing of urban refuse such as garbage, excrement, and corpses, and a similar relationship between sewers and cemeteries existed in Roman cities of the first and second centuries CE.

Conclusion

I have explored some structural sanitary issues of the Roman city through textual and archaeological evidence which help increase our understanding of Roman streets, public spaces, and private dwellings in daily life. Archaeological evidence of sanitation and their interpretation shows how sanitary facilities and practices related to hygiene, rubbish and its disposal, as well as the supply of urban environments with water, shaped and ultimately changed urban society. The evidence certainly shows that Romans put a price on almost everything, that profit and money played a role in most aspects of life. This practice resulted in situations – like the example of the maintenance of bakeries – in which the profitable workings of the business were valued over hygiene and sanitation. The examples I have covered are far from exhaustive, but should be sufficient to stress the complexity of understanding sanitary issues in ancient Rome. The Romans seemed to be exposed to conditions that surely must have been challenging for people living under them.

While some of the stimuli of Roman urban settings might be considered offensive to the senses – smells of urine, excrement, vomit, dung, decay, detritus, garbage, filthy water, mixed with fresh produce and baked goods, but also the odour of dead carcasses, burning or decomposing corpses, boiled food, smoke, charcoal, mould, fermenting or rotting fish sauce – they also may have merely been the smells and sights of home, the familiarity of a real Roman city, and of the comforts of Roman civilization itself[13], as opposed to the barbarity of the provinces or the backward ways of the countryside. One can learn a lot from the complex of Roman customs in the ongoing effort to explore the things they valued and how those things did or did not have anything to do with cleanliness and sanitation.

In the Roman city, excrement and urine had monetary values for agricultural uses (human waste as fertilizer) and for other small-scale industries (urine as a cleaning agent in fulleries). We miss the bigger picture of Roman life if we simply conclude that ancient Rome was a place of poor sanitation without exploring the facts more carefully.

The ancient Roman city of the first and second century CE would have been a very smelly, noisy, unhealthy, and somewhat confusing place, in the eyes of people used to Western sanitary standards and evaluations. The archaeological evidence and testimonies from our literary and epigraphic sources have given us a vivid impression of how all

---

13 I am grateful to Morley (2015: p. 119) for first challenging the notion of what was or was not intrinsically unpleasant or pleasant, threatening or safe, and unsanitary or sanitary in the Roman empire.

five senses were engaged in relation to key institutions and structures of Roman city life, such as the sewer system, water supply, or leisure activities at the public baths and the Colosseum. These sensual experiences can teach us a lot about how Roman society was organized, how cities themselves functioned, and what the quality of life was like for people of different social statuses living in those cities.

We must also appreciate the different perspectives of the Romans on the very meaning of sanitation. The concept of sanitation, hygiene, and cleanliness does come to mean different things in different times and places. The tenacious institutions and structures of Roman urban society were not easily changed, but the Roman legacy of urban life still has a lot to offer contemporary urban planners facing problems related to sanitation today.

Postscript

I am very grateful to the "Clean Team," Isabel Bredenbröker, Christina Hanzen, and Felix Kotzur, in the Research Training School on "Value and Equivalence" at Goethe-Universität, Frankfurt am Main, for having invited me to their conference, "Cleaning and Equivalence," on June 8-11, 2017. The other speakers and their papers were extremely inspiring to me, and gave me much to consider for my own work on Roman sanitation and its value in the ancient Roman city. I also owe the warmest thanks to my husband and fellow traveling companion, Professor Steven E. Ostrow, recently retired from the History Faculty at MIT, for accompanying me on this adventure. We both shall never forget the kindness and spirit (along with the good food and hospitality) we experienced in Germany. We even came away from the conference with a piece of floor dirt embedded inside a crystalized plastic casing which will always remind us that we all leave behind dirt and detritus. My thanks also go to the anonymous reader of this article whose sharp eye, excellent edits, and wide array of suggestions for improvements saved me from presenting various errors and confusions to my readers. Any remaining flaws are my responsibility.

Bibliography

Adams, Tim (2018) London's fatberg on show: 'We thought of pickling it'. *The Guardian.* [Online] Feb. 4, 2018. Available from: https://www.theguardian.com/culture/2018/feb/04/fatberg-museum-london-display-pickling-age-waste.

Aguilera Martin, A. (2002) *El Monte Testaccio y la llanura subaventina. Topografía extra portam Trigeminam.* Rom, Consejo Superior de Ivestigaciones Científicas. Escuela Española de Historia y Archeología en Roma.

Aldrete, G. S. (2006) *Floods of the Tiber in Ancient Rome.* Ancient Society and History. Baltimore, The Johns Hopkins University Press.

Bauer, H. (1989) Die Cloaca Maxima. In: *Vorträge zur Tagung in Rom am 10./11. Oktober 1988. Wassernutzung und Wasserbau im antiken Italien.* Mitteilungen. Leichtweiss-Institut für Wasserbau der Technischen Universität Braunschweig 103. Braunschweig, Leichtweiss-Institut für Wasserbau, pp. 43-67.

Bauer, H. (1993) Cloaca, Cloaca Maxima. In: E. M. Steinby (ed.) *Lexicon Topographicum Urbis Romae. Volume Primo (A-C).* Rom, Edizioni Quasar, pp. 288-290.

Blásquez Martínez, J. M. & Remesal Rodríguez, J. (eds.) (1999) *Estudios sobre el Monte Testaccio (Roma). Vol 1.* Barcelona, Universität de Barcelona.

Blásquez Martínez, J. M. & Remesal Rodríguez, J. (eds.) (2001) *Estudios sobre el Monte Testaccio (Roma). Vol 2.* Barcelona, Universität de Barcelona.

Blásquez Martínez, J. M. & Remesal Rodríguez, J. (eds.) (2003) *Estudios sobre el Monte Testaccio (Roma). Vol 3.* Barcelona, Universität de Barcelona.

Bodel, J. (2000) Dealing with the Dead. In: Hope, V. M. & Marshall, E. (eds.) *Death and Disease in the Ancient City*. London, Routledge, pp. 128-151.

Bomgardner, D. L. (2002) *The Story of the Roman Amphitheatre*. 3rd ed. London, Routledge.

Bradley, M. (2002) 'It all comes out in the wash': Looking harder at the Roman fullonica. *Journal of Roman Archaeology*, 15, 21-44.

Bradley, M. (ed.) (2012) *Rome, Pollution and Propriety. Dirt, Disease and Hygiene in the Eternal City from Antiquity to Modernity*. British School at Rome Studies. Cambridge, Cambridge University Press.

Bradley, M. (ed.) (2015) *Smell and the Ancient Senses. The Senses in Antiquity*. Abingdon, Oxfordshire, Routledge.

Camardo, D. (2011) Case Study: Ercolano: La riscostruzione dei sistemi fognari. In: Jansen, G. C. M., Koloski-Ostrow, A. O. & Moormann, E. M. (eds.) *Roman Toilets. Their Archaeology and Cultural History*. BABESCH. Annual Papers on Mediterranean Archaeology, Supplement 19. Leuven, Peeters, pp. 90-93.

Claridge, A. (2010) *Rome. An Oxford Archaeological Guide*. 2nd ed. Oxford, Oxford University Press.

Clements, A. (2015) Divine Scents and Presence. In: Bradley (ed.) *Smell and the Ancient Senses. The Senses in Antiquity*. Abingdon, Oxfordshire, Routledge, pp. 46-59.

Corbin, A. (1986) *The Foul and the Fragrant. Odor and the French Social Imagination*. Leamington Spa, Berg.

De Caro, S. & Greco, A. (1981) *Campania*. Guide archeologiche Laterza 10. Roma-Bari, Gius. Laterza & Figli.

De Vos, A. & De Vos M. (1982) *Pompei, Ercolano, Stabia*. Guide archeologiche Laterza 11. Roma-Bari, Gius. Laterza & Figli.

Ellis, S. J. R. (2004) The Distribution of Bars at Pompeii. Archaeological, Spatial and Viewshed Analyses. *Journal of Roman Archaeology*, 17, 371-384.

Emmerson, A. (forthcoming) *Life and Death in the Roman Suburb*, esp. chapter 2, "Waste Management in Suburb and Center". Oxford, Oxford University Press.

Fishbein, R. (2016) Toilet Snakes. They Can Happen here too. *Gothamist*. [Online] May 26, 2016. Available from: http://gothamist.com/2016/05/26/toilet_snakes_coming_for_you.php [Accessed: unknown (eds.)]

Flohr, M. (2003) *Fullones* and Roman Society. A Reconsideration. *Journal of Roman Archaeology*, 16, 447-450.

Flohr, M. (2007) Nec quicquam ingenuum habere potest officina? Spatial contexts of urban production at Pompeii, AD 79. *BABESCH. Annual Papers on Mediterranean Archaeology, 82* (1), 129-148.

Flohr, M. (2006) Organizing the Workshop. Water Management in Roman Fullonicae. In: Wiplinger, G. (ed.) *Cura Aquarum in Ephesus. Volume 1. Proceedings of the Twelfth International Congress on the History of Water Management and Hydraulic Engineering in the Mediterranean Region. Ephesus/Selçuk, Turkey, October 2-10, 2004*. BABESCH. Annual Papers on Mediterranean Archaeology, Supplement 12. Leuven, Peeters, pp. 193-199.

Flohr, M. (2013) *The World of the Fullo. Work, Economy, and Society in Roman Italy*. Oxford Studies on the Roman Economy. Oxford, Oxford University Press.

Gee, A. (2016) Which animals can swim up your toilet?. *BBCNews*. [Online] 28 May 2016. Available from: https://www.bbc.com/news/magazine-36397505 [Accessed: unknown (eds.)].

Giordano, C. & Casale, A. (2007) *Profumi, Unguenti e Acconciature in Pompei Antica*. 2nd ed. Revised and updated by Laurentino García y García. Rom, Bardi Editore.

Hobson, B. (2009) *Latrinae et Foricae. Toilets in the Roman World*. London, Duckworth.

Hope, V. M. (2009) *Roman Death. The Dying and the Dead in Ancient Rome*. London, Continuum.

Hope, V. M. & Marshall, E. (eds.) (2000) *Death and Disease in the Ancient City*. London, Routledge.

Hopkins, J. (2012) The 'sacred sewer'. Tradition and Religion in the Cloaca Maxima. In: Bradley, M. (ed.) *Rome, Pollution and Propriety. Dirt, Disease and Hygiene in the Eternal City from Antiquity to Modernity*. British School at Rome Studies. Cambridge, Cambridge University Press, pp. 81-102.

InspectAPedia (2011) *Diagnosing and Curing Sewer or Septic Odors Caused by Building Drains*. [Online] Available from: https://c.ymcdn.com/sites/rfmaonline.site-ym.com/resource/resmgr/crfp/diagnosingandcuringsewerorse.pdf [Accessed: unknown (eds.)]

Jansen, G. C. M. (1995) Der Wasserhaushalt in Ostia Antica. Die ersten Ergebnisse. *Schriftenreihe der Frontinus-Gesellschaft*, 19, 111-123.

Jansen, G. C. M. (2002) *Water in de Romeinse stad Pompeiji – Herculaneum – Ostia*. Leuven, Peeters.

Jansen, G. C. M. (2011) Interpreting Images and Epigraphic Testimony. In: Jansen, G. C. M., Koloski-Ostrow, A. O. & Moormann, E. M. (eds.) *Roman Toilets. Their Archaeology and Cultural History*. BABESCH. Annual Papers on Mediterranean Archaeology, Supplement 19. Leuven, Peeters, pp. 165-176.

Jansen, G. C. M., Koloski-Ostrow, A. O. & Moormann, E. M. (eds.) (2011) *Roman Toilets. Their Archaeology and Cultural History*. BABESCH. Annual Papers on Mediterranean Archaeology, Supplement 19. Leuven, Peeters.

Knowlton, W. T. (1932) Explosions in Sewers and Sewer Maintenance. *Sewage Works Journal*, 4 (1), 114-118.

Koloski-Ostrow, A. O. (2015) Roman Urban Smells. The Archaeological Evidence. In: Bradley, M. (ed.) *Smell and the Ancient Senses*. The Senses in Antiquity. Abingdon, Oxfordshire, Routledge, pp. 90-109.

Koloski-Ostrow, A. O. (2015) *The Archaeology of Sanitation in Roman Italy. Toilets, Sewers, and Water Supply*. Studies in the History of Greece and Rome. Chapel Hill, University of North Carolina Press.

Lanciani, R. (1890) *Ancient Rome in the Light of Recent Discoveries*. Boston, Houghton, Mifflin and Company.

Laurence, R. (1994) *Roman Pompeii. Space and Society*. London, Routledge.

Lemoine, H. (1929-1930) Les Égouts de Paris du XIVe siècle à 1925. *Revue mensuelle de la Chambre syndicale des entrepreneurs de maçonnerie, ciments et béton armé de la ville de Paris et du département de la Seine*, 1929-1930, 4.

Lugli, G. (1934) *I Monumenti antichi di Roma e suburbia*. II. Le grandi opera pubbliche. Rom, G. Bardi.

Maiuri, A. (1958) *Ercolano. I nuovi scavi* (1927-1958) I. Rom, Istituto Poligrafico dello Stato.

McGinn, T. A. J. (2002) Pompeian Brothels and Social History. *Journal of Roman Archaeology* Supplement 47, 7-46.

Meiggs, R. (1960) *Roman Ostia*. Oxford, Oxford University Press.

Mignone, L. M. (2016) *The Republican Aventine and Rome's Social Order*. Ann Arbor, University of Michigan Press.

Morley, N. (2015) Urban Smells and Roman Noses. In: Bradley, M. (ed.) *Smell and the Ancient Senses*. The Senses in Antiquity. Abingdon, Oxfordshire, Routledge, pp. 110-119.

Packer, J. E. (1972) La casa di via Giulio Romano. *Bullettino della Commissione Archaeologica Comunale di Roma*, 81 (1968-1969), 127-148.

Pesando, F. & Guidobaldi, M. P. (2006) *Pompei Oplontis Ercolano Stabiae*. Guide archeologiche Laterza 14. Roma-Bari, Gius. Laterza & Figli.

Poehler, E. E. (2012) The Drainage System at Pompeii. Mechanisms, Operation and Design. *Journal of Roman Archaeology*, 25, 95-120.

Popielarz, T. (2015) Why Manhole Explosions Happen in the Summer. *nbcnews*. [Online] Aug. 19, 2015. Available from: https://www.nbcnews.com/business/energy/flying-manhole-covers-n411921 [Accessed: unknown (eds.)]

Purcell, N. (1996) Rome and the Management of Water. Environment, Culture, and Power. In: Shipley, G. & Salmon, J. (eds.) *Human Landscapes in Classical Antiquity. Environment and Culture*. Leicester-Nottingham Studies in Ancient Society 6. London, Routledge, pp. 180-212.

Reid, D. (1991) *Paris Sewers and Sewermen. Realities and Representations*. Cambridge, Massachusetts, Harvard University Press.

Reimers, P. (1989) 'Opus Omnium Dictu Maximum' Literary Sources for the Knowledge of Roman City Drainage. *Opuscula Romana*, XVII, 137-141.

Reimers, P. (1991) Roman Sewers and Sewerage Networks. Neglected Areas of Study. In: Leander Touati, A.-M., Rystedt, E. & Wikander, Ö. (eds.) *Munuscula Romana. Papers Read at a Conference in Lund (October 1-2, 1988) in Celebration of the Re-opening of the Swedish Institute in Rome*. Skrifter utgvina av Svenska Institut i Rom 8, 17. Stockholm, Åström 111-116.

Rüpke, J. (ed.) (2011) *A Companion to Roman Religion*. Chichester, Wiley-Blackwell.

Scobie, A. (1986) Slums, Sanitation, and Mortality in the Roman World. *Klio. Beiträge zur Alten Geschichte*, 68 (2), 399-433.

Shelton, J.-A. (1998) *As the Romans Did. A Sourcebook in Roman Social History*. New York, Oxford University Press.

Toynbee, J. M. C. (1971) *Death and Burial in the Roman World*. London, Thames and Hudson.

Wallace-Hadrill, A. (2011) *Herculaneum. Past and Future*. London, Frances Lincoln.

Welch, K. E. (2009) *The Roman Amphitheatre. From Its Origins to the Colosseum*. 2nd ed. Cambridge, Cambridge University Press.

Wilson, A. (2003) The Archaeology of the Roman *fullonica*. *Journal of Roman Archaeology*, 16, 442-446.

Wilson, A. & Flohr. M. (2011) The Economy of Ordure. In: Jansen, G. C. M., Koloski-Ostrow, A. O. & Moormann, E. M. (eds.) *Roman Toilets. Their Archaeology and Cultural History*. BABESCH. Annual Papers on Mediterranean Archaeology, Supplement 19. Leuven, Peeters, 147-156.

Ancient sources

Aelian (1959) *On the Characteristics of Animals*. Translated by A. F. Scholfield. Loeb. London: William Heinemann LTD and Cambridge, Massachusetts: Harvard University Press.

Apuleius (1989) *Metamorphoses*. Edited and translated b J. Arthur Hanson. Loeb. London, England and Cambridge, Massachusetts: Harvard University Press.

Cassius Dio (1968) *Dio's Roman History*. Translated by Earnest Cary based on the version of Herbert Baldwin Foster, Vol. III, Loeb. Cambridge, Massachusetts: Harvard University Press.

Cato and Varro (1935) *Cato and Varro*. Translated by William Davis Hooper with revised edition by Harrison Boyd Ash. Loeb. London, England and Cambridge, Massachusetts: Harvard University Press.

Celsus (1935) *De Medicina*. Vol. I. Translated by W. G. Spencer. London: William Heinemann LTD and Cambridge, Massachusetts: Harvard University Press.

Dagron, Gilbert (1918) *Vie et miracles de sainte Thècle*: Texte grec, traduction et commentaire. Brussels.

Horace (1978) *Satires, Epistles, and Ars Poetica*. Translated by H. Rushton Fairclough. London: William Heinemann LTD and Cambridge, Massachusetts: Harvard University Press.

Lucius Junius Moderatus Columella (1941) *On Agriculture*. Translated with a recension of the text by Harrison Boyd Ash. Loeb. London: William Heinemann LTD and Cambridge, Massachusetts: Harvard University Press.

Martial (1993) *Epigrams*. Translated by D. R. Shackleton Bailey. Loeb. London, England and Cambridge, Massachusetts: Harvard University Press.

Pliny the Younger (1969) *Letters, Books VIII-X, and Panegyricus*. Translated by Betty Radice, Vol. II, Loeb. London: William Heinemann LTD and Cambridge, Massachusetts: Harvard University Press.

Polybius (2011) *The Histories*. Translated by W. R. Paton and revised by Frank W. Walbank and Christian Habicht. London, England and Cambridge, Massachusetts: Harvard University Press.

Suetonius (1997) *Vespasian, Vol. II*. Translated by J. C. Rolfe. Loeb. London, England and Cambridge, Massachusetts: Harvard University Press.

Ulpian (1985) *The Digest of Justinian*. Latin Text edited by Theodor Mommsen with the aid of Paul Krueger and translated by Alan Watson. Vol. IV. Philadelphia: University of Pennsylvania Press.

Section Four

Bodies, Objects and Personal Hygiene

A Matter of Representation – Personal Hygiene in Eastern Zhou-dynasty China (771-256 BCE)

Catrin Kost

Bronze vessels are among the most iconic objects of Zhou-dynasty China, and their role in ritual and funerary contexts, as status markers or as bearers of inscriptions has been discussed extensively.[1] Less attention has been devoted to exploring their importance in quotidian practice, for instance, as washing implements. As such, they have mostly been interpreted as a symbol of male wealth or as signifying the importance of female beauty in a society in which marriage alliances between lineages were an important means of maintaining social cohesion. By taking a fresh look at archaeologically excavated materials, this paper proposes that personal hygiene in Zhou China was more than these interpretations suggest.

The following pages outline and discuss selected examples from a group of 26 bronzes dating from the 7/6th to the 4th/3rd centuries BCE, *i.e.* the Spring and Autumn (771-453/403 BCE) and the subsequent Warring States period (453/403-221 BCE). Morphologically, these vessels belong to long-established types. However, the specific terminology in their inscriptions assigned a new meaning to them: they were denoted as implements of personal hygiene.

Although bronze vessels were widely used among the elites of the Zhou realm, current archaeological evidence suggests that the occurrence of such designated washing implements was spatially confined. The related inscriptions and archaeological contexts further attest that these 'washing bronzes' were mainly commissioned by men and for their personal usage. This poses questions about the underlying motivations of local agents and, consequently, the emergence and distribution of the objects has to be viewed and interpreted within the broader context of socio-political transformations.

1 The author would like to thank Prof. Zhu Fenghan (Peking University, Department of History) and Ondřej Škrabal, M.A. (Charles University, Prague), for their help in introducing me to bronzes and bronze inscriptions. I am grateful to Dr. Shumon Hussain (Aarhus University) and PD Dr. Maria Khayutina (Munich University) for their willingness to discuss the ideas put forth here. An anonymous reviewer provided helpful comments on an earlier draft of this paper. Naturally, all remaining errors are my own.

in: Bredenbröker, I., Hanzen, C. & Kotzur, F (eds) 2020: *Cleaning and Value. Interdisciplinary Investigations*, Sidestone Press (Leiden), pp. 129-154.

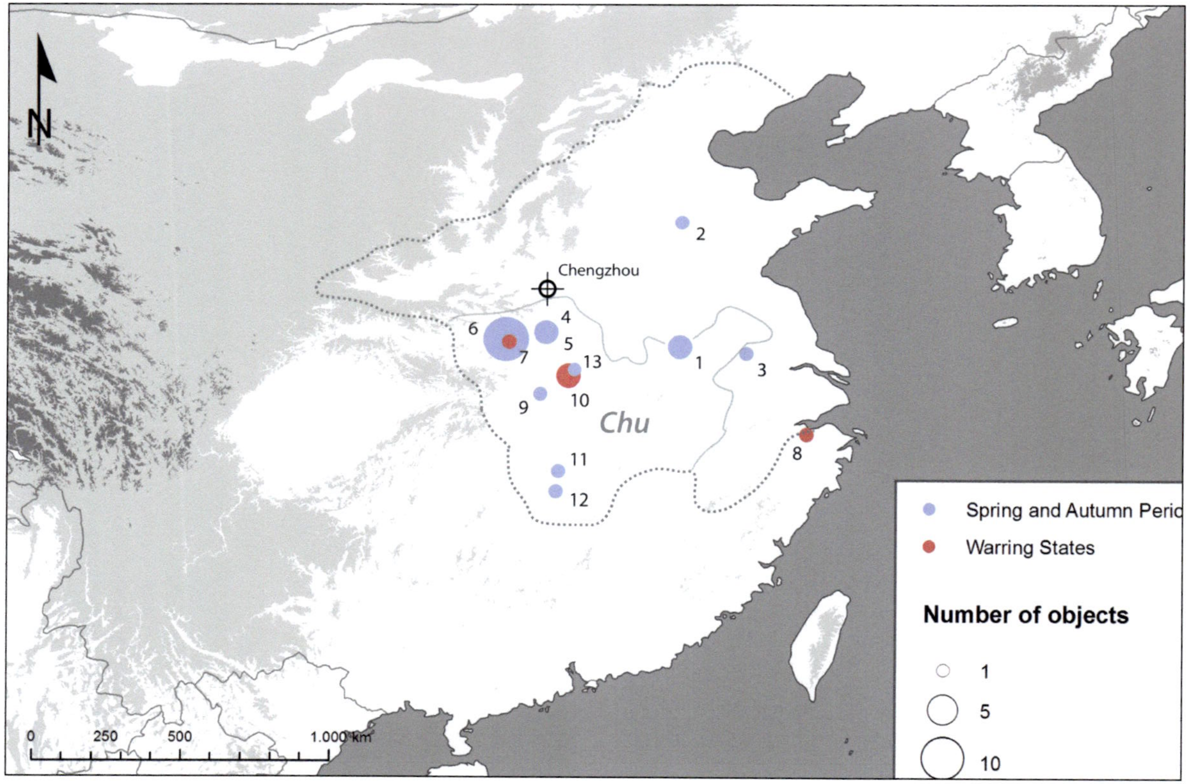

Figure 1. Approximate extent of the Zhou realm and the state of Chu in the 6th cent. BCE. Also indicated are relevant archaeological findspots and the historical period they date to. Washing vessels stem from the following sites: 1. Shouxian 2. Xianrentai 3. Chengqiao Middle School 4. Wancheng 5. Wuzicheng 6. Xiasi 7. Xujialing 8. Shizishan West 9. Huangtupo 10. Wenfengta 11. Gaoquanshan 12. Yangjiashan 13. Yidigang (general map based upon Gernet, 1998: map 3; location of sites determined with the help of GIS data).

I argue that larger trajectories of change starting in the late Western Zhou period (ca. 9th/8th cent. BCE) provided a suitable framework for the emergence of this 'washing' phenomenon: due to the waning power of the Zhou kings, this time saw a loosening of the ties between centre and periphery. One consequence of this development was the growing independence and interaction of different satellite states, who would eventually start competing with each other in an attempt to increase both their political and territorial influence. The ensuing struggles leading to the Warring States period, however, were not only fought by force of arms but also through what we may call 'intellectual' warfare; in their attempt to navigate a highly volatile and constantly shifting social framework, contemporary political agents developed creative means of displaying their superiority. In this climate, practices of personal hygiene gained a novel significance and the conscious display of objects associated with them – and clearly marked as such – became a means of communication and distinction. The emergence of washing vessels is thus not to be interpreted as merely functional or even 'innocent'; instead, it should be seen as a material expression for a strategy of dealing with a changing socio-cultural framework.

Methodological Approach and Source Materials

By examining the emergence, circulation and usage of washing vessels, this article proposes a new perspective on the relevance of personal hygiene in Eastern Zhou China. The source material scrutinised in the following comprises 26 bronzes vessels excavated from twelve different cemeteries located in the modern-day provinces of Anhui, Shandong, Jiangsu, Henan, Zhejiang, Hubei and Hunan (fig. 1; cf. Appendix). The relevant finds date from the mid-Spring and Autumn period (in the following also Chunqiu period) to the mid/late Warring States period (hereafter also Zhanguo period), which corresponds to the time from the 7th/6th to the 4th/3rd centuries BCE.

With thousands of bronze vessels hitherto known for this period, the small size of the sample requires explanation; only objects from archaeologically excavated contexts and with inscriptions referring to aspects of washing were selected for the subsequent analysis. The first criterion is meant to secure a maximum amount of contextual information; the second is based on the general persuasion that inscriptions regarding function on individual vessels can (and should) not automatically be extrapolated to all similar vessels (see infra). Instead, we have to acknowledge that even the same type of object may change in significance throughout history and be reinterpreted as time goes by. Inscriptions of a later period may then, for instance, not document a change in usage but rather point to a new significance of particular object-types that is tied to the specific place of these artefacts in an altered socio-historical context; inferring the same 'function' or 'role' for their predecessors would thus be methodologically problematic.

The present examination, therefore, assumes that the *act* of inscribing and thus emphasising or allocating a certain purpose to objects is historically significant in itself. Following this logic, one at least has to ask the question why particular bronze vessels were deliberately designated as washing implements at a particular point in time (and not before), rather than simply assuming that the related inscriptions indicate that similar/identical usages were already in place at earlier times. The point here is not so much that this assumption is generally incorrect, but rather that it feigns the view for the possibility that the act of 'labelling' may mark a new 'stage' on the long-term trajectory of an object's functional crystallisation.

A conservative approach such as the one adopted here thus requires the inspection of the inventory of unambiguously washing-related objects for which a maximum amount of contextual information is available. Twenty-seven bronze vessels meeting these requirements were compiled both from the data available through the database *Digital Archives of Bronze Images and Inscriptions*[2] and from recent archaeological literature. In selecting objects with relevant inscriptions, I follow the transcription and interpretation of characters proposed by the respective authors as well as the scholars contributing to the database. General information about provenance, dating, gender of the grave owner, vessel type, and basic content of inscription(s) have been tabulated in the Appendix. These data form the basis for the subsequent reflections.

The perspective assumed here is interpretive. My goal is not only to confront the newly derived data with previous narratives about washing practices and the historical meaning of the respective vessel types, but to develop a contextual understanding of inscribed and thus designated 'washing vessels' at a point of critical socio-historical transition, *i.e.* the shift from the Spring and Autumn to the Warring States period. How do these objects and their inscriptions resonate with other developments of their time? What, if any, is the significance of explicitly inscribing object-types that were already in place for a long time with a novel meaning?

In order to approach these issues and to suggest some possible solutions, the paper proceeds in three steps. Following a brief overview of the general role of bronze vessels in Zhou society, I outline the specific terminology employed to distinguish washing vessels from other bronzes before scrutinising some relevant inscriptions more closely. These provide some critical insights and, in conjunction with other examined inscriptions, allow for inferences on the gender, status and identity of those commissioning and receiving the vessels; the inscriptions are also informative about the social groups within which 'washing bronzes' most likely played a crucial role. Together with the geographic distribution of these objects, this overall configuration is then placed into its larger historical context, highlighting parallel transformations which affected other domains of society and which may have imbued washing practices with unprecedented importance.

Bronze Vessels and their Role in Zhou Society

The bronze casting industry of the Zhou 周 dynasty (1046-256 BCE)[3] produced a wide variety of bronze vessels. While bronzes of the Western Zhou 西周 (1046-771 BCE; cf. *e.g.* Rawson, 1990) initially continued artistic traditions of the preceding Shang 商 dynasty (1600-1046 BCE; cf., *e.g.* Bagley, 1987), they were soon complemented with a new stylistic vocabulary (cf., *e.g.* So, 1995). Apart from intricate decorations, the objects could also bear inscriptions, whose contents ranged from simple clan signs

2 The *Digital Archives of Bronze Images and Inscriptions* 殷周金文暨青銅器資料庫, a database established and maintained by the Academia Sinica in Taipei, lists a far larger number of washing vessels, which also comprises museum objects. Due to missing information about provenance and/or archaeological context, these objects do not form part of the following analysis. For further information on these objects, cf. http://www.ihp.sinica.edu.tw/~bronze/; last accessed 28-March 2018.

3 The period of the Zhou Dynasty comprises the earlier Western Zhou (Xi Zhou 西周; 1046-771 BCE) and the subsequent Eastern Zhou (Dong Zhou 東周; 771-256 BCE), which ended with the deposition of the last Zhou king. Commonly, the Eastern Zhou is divided into two periods, referred to as Spring and Autumn period (Chunqiu 春秋; 771-453/403 BCE) and Warring States period (Zhanguo 戰國; 453/403-221 BCE). Throughout the text, mainly the latter denominations are used.

to elaborate texts commemorating specific events and communicating political and social achievements of their owners (see, *e.g.* Qiu, 2000:pp.64ff.). Bronzes can thus be located at the intersection of materiality and textuality and contribute to a better understanding of Zhou history (*e.g.* Shaughnessy, 1991).

Bronze vessels had both a functional and a symbolic role. Traces of liquids, grains and bones found in excavated objects (cf. *e.g.* Okamura, 2005) are proof for their utilisation as containers; however, they were also key implements in ritual contexts, which took the form of feasts in ancestral temples, to which 'the ancestors were thought to descend' (von Falkenhausen, 2006:p.47). From the late Western Zhou dynasty on, vessels also occur in fixed numbers as parts of funerary sets, their quantity and quality serving as an indicator for the social status of the individuals buried with them (see *e.g.* von Falkenhausen, 1999:pp.470ff; von Falkenhausen, 2006:pp.293-325; Rawson, 1999:pp.433ff).

Production and circulation of bronzes seem to have been restricted mainly to the social elite of the Zhou (*e.g.* Yuan, 2009)[4] and the objects are thus interpreted as symbols of power, and as a means of both embodying and installing social relations (Sun, 2010; Cook, 1997); that the vessels were highly prised is corroborated by the fact that they were often handed down across generations within the same lineage before actually being interred (Zhu, 2009:pp.1264-1265).

The study of bronzes and bronze inscriptions thus provides insight into the socio-cultural framework of Zhou-dynasty China and allows for a more detailed knowledge of historical events, social relations and religious practices as well as changes therein.

Bronze Vessels as Washing Vessels

Zhou-dynasty bronze vessels come in an astonishing array of shapes and are commonly assigned to three broader categories – namely food, liquor and water vessels – which can be further distinguished; liquor vessels are often subdivided into vessels for warming, storing or drinking liquor (*e.g.* Zhu, 2009), while food vessels are sometimes differentiated as meat or grain offering containers (*e.g.* von Falkenhausen, 2001:p.220). This categorisation, albeit traditional[5], is somewhat idealised; it assumes that functions inferred from inscriptions on individual vessels can be extended to *all* vessels of similar shapes, thereby establishing a universal, ahistorical functional typology.

The 26 washing bronzes transgress theses 'classical' categories. They comprise *fou* 缶-flasks, *ding* 鼎-tripods[6], *pan* 盤*-basins and yi* 匜*-ewers (fig. 2), dou* 斗*-ladles and thus vessel types* – according to their functional morphology – would have served as water/liquor ladle

4 It should be noted, though, that at the Zhouyuan site, bronze vessels have also been found in graves of people thought to be involved in the casting process (Shaanxisheng kaogu yanjiuyuan, 2010). This raises questions as to the precise nature of the Zhou-dynasty economy as well as the relationship between craft production and social change. While a more detailed discussion of the first subject is still outstanding, the second topic is examined in, e.g., Ma, 2016.

5 This division is still closely related to the typology first set out in the 'Illustrated Catalogue of Examined Antiquity' (*Kaogutu* 考古圖). The work by Lü Dalin (呂大臨; 1046-1092 CE), among other things, systematically describes and classifies bronzes from the imperial as well as private collections (see Moser, 2012 for a novel interpretation of the work, including further literature). The slightly varying approaches to classification made by Chinese scholars of the 20th century are nicely subsumed in Zhu, 2009:pp.78-83.

6 While *ding*-cauldrons are usually addressed as tripods, they also occur in a quadripod form.

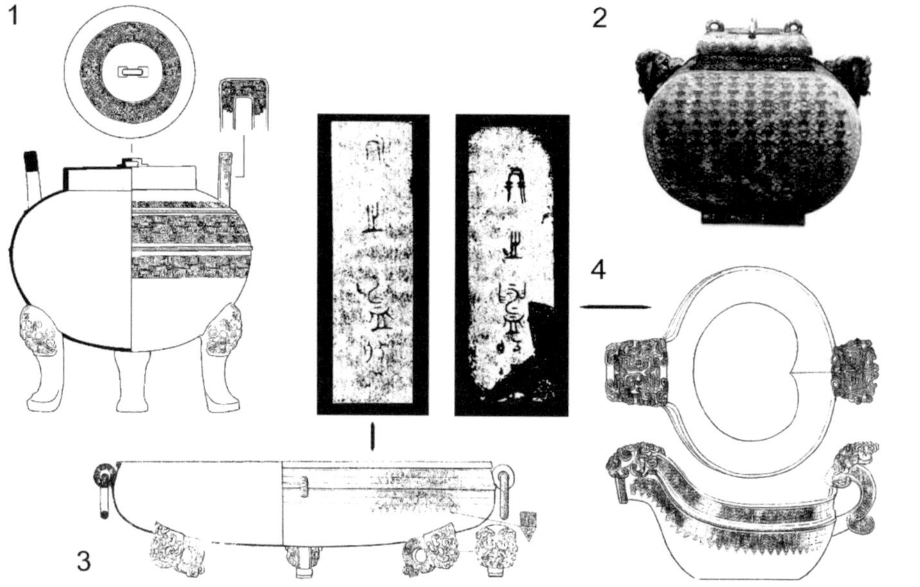

Figure 2. Overview of different vessel types denominated as washing implements. 1. *Ding*-tripod, Wancheng grave M38, Henan; 2. *Fou*-flask, Wenfengta grave M18, Hubei; 3. *Pan*-basin, Xiasi grave M2; Henan; 4. *Yi*-ewer, Xiasi grave M2. (1. After Nanyangshi wenwu kaogu yanjiusuo, 2011: fig. 16; 2. Based on Hubeisheng wenwu kaogu yanjiusuo & Suizhoushi bowuguan, 2014a: fig. 27; 3. & 4. After Zhong et al, 2006:p.334).

(*dou*) or containers (*fou*)[7], cooking and/or serving vessels (*ding*), and water vessels (*pan*; *yi*).[8] Furthermore, all five vessel shapes can be traced back to earlier periods and were already well-established at the beginning of the Zhou Dynasty.[9] It is not before several hundred years later, however, that they are labelled as instruments of personal hygiene.

We therefore have to acknowledge the possibility that the vessels were utilised in different contexts than those related to personal hygiene. However, as outlined above, the act of designating them as washing implements might already be meaningful in itself and can be interpreted as signalling a changed attitude towards a class of objects and the practices to which they were connected.

7 Note that there are two different types of *fou*-flasks, for which written records suggest a different usage (cf. e.g. Zhu, 2009:pp.221-224). *Fou* (sometimes also *yu fou* 浴缶) were used as water, *zun fou* 尊缶 as wine containers (Zhu, 2009:pp.221-222).

8 For a good overview of the usage and designation of *dou*, *ding*, *fou*, *pan* and *yi* as we can understand it through bronze inscriptions, written sources and archaeological finds, see Zhu, 2009:pp.86-112, 221-224, 268-273, 279-288, 288-295, .

9 For the sake of completeness, it should be mentioned that, although the shape of *fou*-flasks has antecedents, the denomination as such does not occur before the mid/late Chunqiu period (Zhu, 2009:p.221). Before that time, flasks were usually addressed as *hu* 壶 (cf. e.g. Zhu, 2009:pp.224-243).

Zhou-dynasty washing terminology and inscriptions

'Clear' characters? Remarks on specific terms and related problems

Particular expressions for washing practices appear for the first time in somewhat greater numbers during the latter part of the Zhou dynasty, with the earliest example dating to the mid-Chunqiu period, *i.e.* the 7th/6th century BCE (cf. Appendix).[10] The washing terminology of this time comprises six different expressions, which – following the explanations given in Xu Shen's 許慎 (ca. 55-149 CE) etymological glossary *Shuowen jiezi* 說文解字 (lit. 'Explaining graphs and analysing characters') can be translated in the following way:

盥 *guan* as 'to wash one's hands', 沬 *mei* as 'to wash one's face', whereas 沐 *mu* means 'to wash one's hair' and 浴 *yu* 'to wash one's body'. While these four characters refer to washing practices directed at specific parts of the body, 湯 *tang* and 浣 *huan* are more generic and simply describe the act of 'bathing' or 'soaking oneself in water' (*tang*)[11] as well as 'washing oneself' (*huan*).

In listing these seemingly unambiguous translations it has to be stressed that the *Shuowen* is of a much more recent date than the objects in focus here – the original version of the work, although largely based on earlier materials, was only finished around 100 CE.[12] The subsequent loss of large parts of the work and a long history of editing further contribute to make the *Shuowen* a rather sensitive source (cf. Boltz, 1993) – one that is, however, indispensable.

If we broadly accept the 'washing terminology' outlined above despite these uncertainties, we can make the following observations about the employed terms based on the 26 examined bronze objects:

Different characters distinguish various hygienic practices that were, at least for the most part, directed at different parts of the body, such as hands, face etc. Although the explanation of their meaning stems from a much later source, the presence of the respective characters at this earlier stage is certainly indicative of developments leading to the 'evolved' washing terminology in the *Shuowen*.

The characters occur with different frequencies: *guan* (10), *yu* (5), *tang* (4) and *mu* (4) are more commonly used; *mei* (1) and *huan* (2) appear more seldom.

The chronology/dating of the various vessels on which the characters are attested indicates that most of the terms were prevalent at roughly the same time – the late Chunqiu period (ca. 6/5th cent. BCE) (cf. Appendix). Only one character forms an exception from this rule: *guan* is already evidenced in an inscription dating to the mid-Chunqiu period (7th/6th cent. BCE; Zhang, Hu & Xiang, 2004:p.34). Although the number of analysed objects is fairly moderate (n=26), this temporal trend is certainly noteworthy.

The geographic distribution and focus of the various terms differ: the character *guan*, for instance, is evidenced through finds from the modern-day provinces of Hunan, Hubei, Jiangsu, Shandong, Henan and Anhui (fig. 1, sites no. 1; 2; 3; 6; 9; 11) and thus in a larger

---

10 Some authors (Yin & Ren, 2003:p.2; Liu, 1998:p.1) have voiced the opinion that part of the washing terms discussed in the following can be traced back to the times of the Shang-dynasty (1766-1122 BCE), during which they are thought to have occurred in the so-called oracle bone inscriptions. The argument as well as the related idea of an uninterrupted usage of characters in the very same sense, however, is not very convincing.

11 Li Ling (1991) translates tang 湯 as 'hot water'.

12 For a discussion of the production of the text and its place in the history of scientific thought see Bottéro & Harbsmeier, 2008.

area, whereas the character *yu* is solely known from inscriptions unearthed in modern-day Henan Province (fig. 1, sites no. 6; 4). The problem of incomplete discovery must surely be considered here; yet, the overall spatial trend may be of historical significance since the pattern is consistent with and 'makes sense' in the light of contextual arguments derived from other sources of evidence (see *infra*).

At a first glance, it may seem tempting to think about the importance of different washing practices based on the frequency with which they are mentioned in inscriptions. However, instead of speculating whether in the Eastern Zhou period, washing one's hands ranked over washing one's hair, I suggest seeing the quantitative occurrence of these terms in relation to the emergence, or maybe rather 'textualisation', of a more fine-grained washing terminology. This pattern would be consistent with the typical development of 'cultural techniques' (*Kulturtechniken*) over time. These have been shown (cf. contributions in Kassung & Macho, 2013) to have much older origins than the terms and concepts later associated with them. As Thomas Macho (2003:p.179) has prominently put it:

> "Cultural techniques – such as writing, reading, painting, counting, making music – are always older than the concepts that are generated from them. People wrote long before they conceptualised writing or alphabets; millennia passed before pictures and statues gave rise to the concept of the image; and until today, people sing or make music without knowing anything about tones or musical notation systems. Counting, too, is older than the notion of numbers. To be sure, most cultures counted or performed certain mathematical operations; but they did not necessarily derive from this a concept of number."

We can thus assume that not only washing practices but also some of the related expressions had been in use in earlier times and were gradually put into writing from the $7^{th}/6^{th}$ century BCE on. In the following centuries, complementary phrases were added, eventually resulting in a terminology that also illustrates a fine-grained 'body topology'. The climax of this process is clearly reflected in written sources of the Han Dynasty (206 BCE-9 CE; 25-220 CE), which employ a plethora of expressions to differentiate practices of personal hygiene. With the help of these terms, written sources also provide guidelines for the frequency and timing of bodily care (cf. Kost, 2017:pp.147-150).

Some washing terms are employed across a large area, while others suggest that we should acknowledge the possibility of local developments. The character *yu* is hitherto solely known from inscriptions unearthed in modern-day Henan Province, with five relevant finds stemming from the cemetery of Xiasi (Henansheng wenwu yanjiusuo, Henansheng Danjiang ku qu kaogu fa jue dui, Xichuanxian bowuguan, 1991), which has been identified as a cemetery of the Wei 鄬-lineage – a junior lineage of the ruling family of the state of Chu (Henansheng bowuguan, Xichuanxian wenguanhui & Nanyang diqu wenguanhui, 1981:p.126).[13] The expression *yu* as hitherto evidenced through archaeological finds, was thus not only used locally, but also within specific and interrelated social groups.

13 For more information on the cemetery and the lineage see e.g. von Falkenhausen, 2006:pp.338-348. On the identification of the principal occupant of tomb 2 with Yuan Zi Feng 虤子倗, a chief minister (lingyin 令尹) of Chu, see Li, 1981.

We can thus summarize that an increasing presence of washing terms in bronze inscriptions is so far attested mainly for the late Chunqiu period (ca. 6/5th cent. BCE). During this timeframe, some expressions appear to have been well-established and in use throughout large geographical areas, while others were less common or even confined to a local, intergroup usage. Drawing upon the ideas voiced in regards to the development of 'cultural techniques', we can posit that the emergence of a washing terminology apparently 'lagged behind' and that bronze vessels were likely employed as washing implements *before* actually being addressed as such. The occurrence of sets of water vessels supports these ideas. From the early phase of the mid-Western Zhou period on, *pan*-trays and *yi*-ewers are usually are associated in the archaeological record, with the *yi* placed inside the *pan*. This combination, however, was preceded by a set consisting of *pan*-basin and *he* 盉, a three-legged pouring vessel with closed spout. There is a chronological overlap between both sets, which seem to have been functionally identical. But *he*-vessels eventually disappeared from the archaeological record.

From indicating ownership to expressing reverence – inscriptions on washing bronzes

The inscriptions on the 26 washing vessels can be divided into two bigger categories (cf. Appendix). An example for the first group stems from the *yi* 匜-ewer excavated from the grave of She, head of the Peng lineage (fig. 1, site no. 4; fig. 3.3). The object with the bowl-shaped body has a handle rendered in the shape of a dragon with a coiled tail and a large spout whose upper part is embellished with an animal mask. The upper half of the vessel is decorated with two broad bands of a coiled serpent pattern; separated by a raised ridge; the lower band is further combined with a triangular cicada pattern. The inscription is cast into the flat base on the inside of the vessel and consists of seven characters in two lines. It translates as 'Face-washing *yi*-ewer belonging to She, head of the Peng lineage' 彭子射之行會(沫)臾(匜), thus giving a clear indication of both the usage of the vessel as well as the identity of its owner.[14]

Inscriptions of the second type are longer, providing us with more detailed information, for instance on the date and the occasion of the vessel production, the commissioner and the individual it was dedicated to.

A *pan*-basin (fig. 3.2) from Huangtupo grave M3 in Hubei Province (fig. 1, site no. 9; Jingzhou bowuguan & Zhongxiangshi bowuguan, 2009) may serve as an example: the round and slightly curved vessel has three short legs decorated with animal masks and four loops – two of them with an additional ring suspended from them – attached to its body. The decoration covering the upper half of the merely 8.9 cm high object consists of a coiled serpent pattern, a triangular cicada pattern and a twisted cord pattern. The centre of the basin bears an inscription of four lines, which we can translate as follows:

---

14 While the identity of the grave owner has been clearly established, the transcription – and thus also the translation – of the inscription, remain subject to discussion. The English translations given here mostly relies on the work by Cao Jinyan (2011:p.94), who has pointed out that 臾 should be read as 匜 and reads the second to last characters as 会, which he translates as mei 沬, i.e. to wash the face. I did not follow the author's suggestion to translate '行' as 'travelling' and thereby interpret the vessel as an accessory used (exclusively) when going on a journey (Cao Jinyan, 2011:p.94; for a discussion of the term see also Khayutina, 2010: footnote 87). Instead, I am much indebted for the advice given by Professor Zhu Fenghan 朱凤瀚 (Chinese Ancient History Research Centre, Peking University) to understand the term in the sense of 'to use', 'to employ', i.e. an *yi*-ewer *used* for washing the face.

'It was the beginning auspiciousness of the first month on *dinghai* day[15] that the head of the Deng lineage [had made] this hand-washing basin for the middle-born [of the] Man as dowry. May she [live forever/10000 years] without limits, may sons and grandsons forever [treasure [it]].'

唯正月初吉丁亥，鄧子□賸叔曼盥盤，□□無諆，子孫永□

With the help of this text, we learn that during the late Spring and Autumn period, the third-born daughter of the head of the Man-surnamed (cf. Khayutina,2014:p.50) Deng lineage, received this vessel as dowry from her father.[16] The inscription further expresses the wish for her to live a long life, and for the vessel to be treasured by coming generations[17]. Especially this last part is a common phrase in bronze inscriptions, which indicates the reverence with which the objects were viewed, treated, and sometimes handed down as family heirlooms for generations. Of the 25 inscriptions analysed here, 12 give expression to this desire to be used and appreciated – either by descendants or the commissioners themselves (cf. Appendix). The remaining 14, however, mainly state ownership in the style of the previously mentioned inscription on the *yi*-ewer of She, head of the Peng lineage. Given the sampling rationale followed in this study (see 'methodological approach and source materials'), this figure may trace some of the prevailing norms in the period of interest.

Washing vessel inscriptions thereby illustrate a general trend that has been observed when comparing bronze inscriptions from the Western Zhou to those of the later Eastern Zhou (*e.g.* von Falkenhausen, 2006:pp.292ff): whereas earlier inscriptions would state in more detail the circumstances under which a bronze vessel came into being, *e.g.* who commissioned it, why he obtained the privilege to do so and whom the vessel eventually was gifted to, they also explain how the object was to be used in the worship of the ancestors (von Falkenhausen, 2006:pp.294-295). Later bronzes – although still used in ritual contexts – less frequently have this clear focus on the forebears; instead of naming the specific ancestor in whose sacrifices they should be employed, now the inscriptions commonly state that vessels were made for the donor's personal usage[18]. During the Eastern Zhou, the self-representational dimension

---

15 As this inscription indicates, the Zhou noted the months of a year by number and the days by using the sexagenary cycle (also *ganzhi* 干支), made up by a combination of ten 'stems' and twelve 'branches'. They further divided each month into four periods, of which the 'beginning auspiciousness' is the first. For more information on the calendar of the Zhou and a discussion of the four periods as well as their length, see Shaughnessy, 2009.

16 Although clearly intended for the third-born daughter, it is not entirely sure whether the *pan*-basin also remained in her possessions. The inscription itself gives rise to doubts since – probably in an attempt to remove it – it was severely scratched (Jingzhou bowuguan & Zhongxiangshi bowuguan, 2009:p.277). This attempt to erase the association of the vessel with a specific individual lead the excavators to suggest that the tomb owner of the grave in which the *pan*-basin was found might be a different person and that the bronze vessel might have changed hands during the annexation and final overthrow of the state of Deng 鄧 through king Wen of Chu 楚文王 in 678 BCE (Jingzhou bowuguan & Zhongxiangshi bowuguan, 2009:pp.289f.).

17 The mentioning of 'sons and grandsons' 子孫 or 'sons' sons and grandsons' grandsons' 子子孫孫 is ubiquitous in bronze inscriptions and usually translated in this way. For a brief discussion on whether this wish for abundant progeny should be understood as comprising offspring of both sexes, see Khayutina, 2014:p.33.

18 A few bronzes even state clearly that they were used in rituals carried out with regard to the community. See, e.g. the translation of a proportion of the inscription on the Wangsun Gao-*yongzhong* bells from Tomb 2 at the Xiasi cemetery (Henan) given in von Falkenhausen, 2006:p.296. Its content makes clear that – contrary to earlier times – 'these bells were to be played at a ritual banquet celebrating and sanctioning the donor's allegiance to his overlord, rather than at a sacrifice renewing his links with his ancestors'.

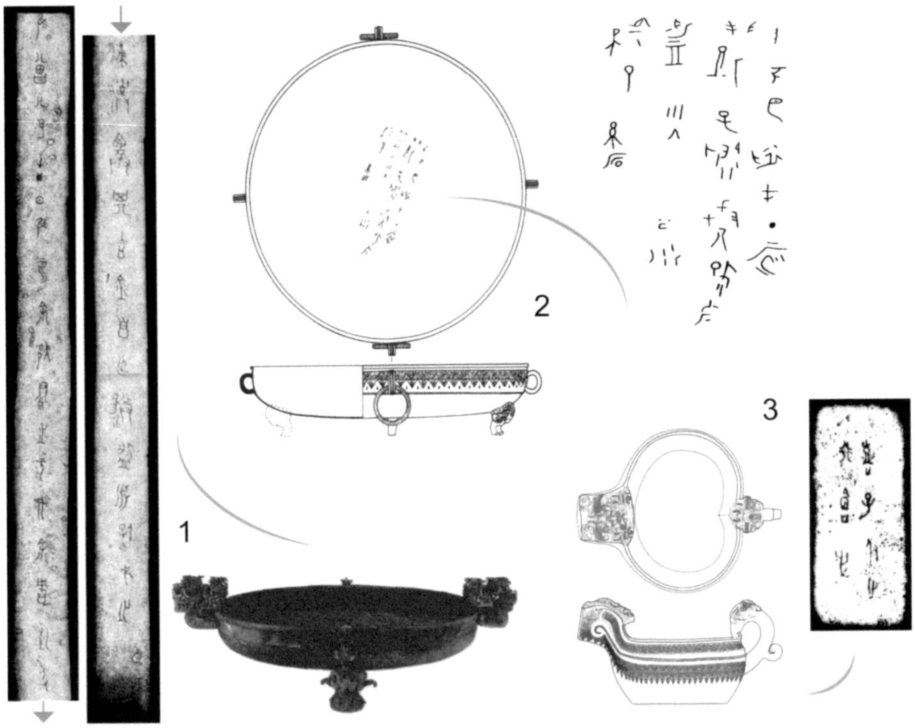

Figure 3. Washing vessels and rubbings of inscriptions on them. 1. *Pan*-basin, Wenfengta grave M33, Hubei; 2. *Pan*-basin, Huangtupo grave M3, Hubei; 3. *Yi*-ewer, Wancheng grave M38, Henan. (1. Based on Hubeisheng wenwu kaogu yanjiusuo & Suizhoushi bowuguan, 2014a: fig. 32; 2. After Jingzhou bowuguan & Zhongxiangshi bowuguan 2009: fig. 30.3; 3. After Nanyangshi wenwu kaogu yanjiusuo, 2011: fig. 51).

of bronze vessels seems to have been emphasised and their role in forging social ties within the community of the living, largely unrelated to practices of ancestral worship, thus gained increasing importance (cf. *e.g.* Rawson, 1990:p.94; Major & Cook, 2017:pp.111f.). Even though bronze objects probably always served some self-representational purpose, this shift in their matrix of significance is clearly meaningful; it is likely tied to more general transformations affecting Zhou societies at the time. That the distinctively social aspects of enacting bronze objects became more and more prominent is at least supported by the presently analysed data set: only two of the 26 bronze objects – a *fou*-flask from Wenfengta grave M18 (Hubeisheng wenwu kaogu yanjiusuo & Suizhoushi bowuguan, 2014a:pp.27f.) and an *yi*-ewer from Yangjiashan (Hunansheng bowuguan, 1963:p.680) – bear inscriptions explicitly mentioning or referring to sacrificial activities (cf. Appendix).[19]

---

19 'Negative' evidence, that is, evidence of absence, should be invoked cautiously. However, since the particular point made above contributes to a broader contextual argument, it can reasonably be considered; the argument does not stand on itself – in which case it would be difficult to defend – but 'resonates' positively or negatively with other lines of evidence (see *infra*).

Type			Vessel			
Commissioner => Dedicatee	*pan*-**basin**	*yi*-**ewer**	*fou*-**flask**	*ding*-**tripod**	*dou*-**ladle**	n =
man => himself	2	5	5	5	1	18
husband => wife	1					1
father => daughter	2		1 (?)			2 (3?)
woman => herself	1		2	1		4
n =	6	5	8	6	1	25 (26?)

Table 1. Gender of commissioners and dedicatees as determined through inscriptions on analysed washing vessels (N = 26).

Whose Vessel and for Whom? Gender, Identity and Status of Commissioners and Dedicatees

The inscriptions discussed above, but also the majority of the 26 examined bronze vessels can be characterised as 'personal objects' – they were intended to be used either by the person who ordered them or the person they were gifted to (cf. Appendix). Whether commissioners or dedicatees, all persons associated with washing vessels were of high status; the head of the Deng lineage and his third-born daughter or aristocrats in the service of the ruling families such as She – they all belonged to the social and political elite of Eastern Zhou-period China. The same holds true for the other persons mentioned in the analysed inscriptions, which we can identify as lineage heads 子 *zi*, lords 侯 *hou*, dukes 公 *gong*[20] etc., some them relatives of the first king of the Zhou Dynasty, King Wu 周武王[21] (r. probably 1049/1045-1043 BCE).

Apart from being personal objects of the members of the Zhou elite, the inscriptions further suggest that washing vessels were gender-specific. Those cases in which the identity of the commissioner could be determined illustrate that a large proportion of the vessels (n=18) were ordered by men and for themselves (table 1; Appendix).

More seldom, male commissioners would gift washing bronzes to their wife (n=1) or daughter (n=2(3?)). Although female commissioners are by no means rare (cf. *e.g.* Khayutina, 2014), the data hitherto available only provides evidence for two women who ordered washing vessels for their own usage (n=4). One of the related objects is the *pan*-basin (fig. 3.1) excavated from grave M33 at the Wenfengta cemetery in Hubei province (Hubeisheng wenwu kaogu yanjiusuo & Suizhoushi bowuguan, 2014a; Hubeisheng wenwu kaogu yanjiusuo & Suizhoushi bowuguan, 2014b). The round basin has three animal-shaped legs and two lavishly decorated, openwork handles[22]. Two dragon-like creatures are attached to its walls, seemingly climbing

20 The *ding*-tripod excavated from Xujialing grave HXXM11 in Henan Province (Wang & Hao, 2009; Wang & Qiao, 2009; Feng, 2009) and the *fou*-flask from Wenfengta grave M61 at first glance seem like exceptions to this rule, as the commissioner of the first vessel is called 'Ying, the great chief in Wei' 大尹(君)嬴, whereas the second is addressed as both 'Qiao, the chief of ... in Zeng' 曾旨(?)尹喬 and, in another inscription, as 'Qiao, commander-in-chief in Zeng' 曾大司馬喬. These titles, however, indicate high-ranking positions within the government, which were mostly given to the members of the royal family (cf. e.g. Zhou, Chen & Qi, 1998:pp.3f.; 37f.).]

21 King Wu overthrew the Shang Dynasty in 1046. BCE during the battle of *Muye* 牧野 and delegated some of his younger brothers to oversee the former Shang domain. For more information on Western Zhou history, see Shaughnessy, 1999.

22 The openwork handles on this vessel are exceptional since they provide proof for the use of the lost-wax-casting technology as early as the mid- to late Spring and Autumn period (Hubeisheng wenwu kaogu yanjiusuo & Suizhoushi bowuguan, 2014a:p.33).

it. Contrary to the *pan*-basin from Huangtupo grave M3, this one is decorated on the inside and with an interlaced serpent- and a dragon pattern. This is also where we find the following inscription running all the way around the inside of the rim:

> It was the 8th month of the *Zeng* [calendar], on the auspicious day of *hai*, that I, Pu, the first-born daughter of the Lord of Fu (?), the wife (?) of the nephew (?) of the Zhou lineage, selected this auspicious metal and had made on my own initiative this washing-basin. (May I?) forever treasure and use it!
>
> 唯曾八月，吉日唯亥，余君之元女，余周侄婦，擇其吉金，自作 (浣)[23]盤，永保用之.

Not all inscriptions on washing vessels mention specific names, but those hitherto known are personal names rather than posthumous titles. This is illustrated through vessels excavated at Shouxian in Anhui province (Chen, 1956) that list the personal name of the grave owner – Shen 申 – instead of his known posthumous title Lord Zhao of Cai 蔡昭侯. From this, we can infer that washing vessels were made for use during life and not for the funerary context itself. A *fou*-flask (Chen, 1956:p.108) from the very same grave, however, shows that they could well be returned to their commissioner, *e.g.* in the course of the burial rites. According to the inscription, the flask was originally commissioned by Lord Shen of Cai for his eldest daughter[24] on the occasion of her marriage. Along with several other vessels, the lady addressed as 'Great Elder Ji' 大孟姬 returned it to him, probably as a funerary gift, when he was interred.

Putting the Pieces together: Loosening Ties and Shifting Identities

The combined evidence of 'washing' vessels presented so far permits the formulation of the following four hypotheses: (i) although a few earlier (7/6th cent. BCE) and later (4th/3rd cent. BCE) examples are known, specifically designated washing vessels are mainly a phenomenon of the 6/5th century BCE. The fact that certain 'sets' (*pan* and *he*; *pan* and *yi*) likely associated with washing practices (see *e.g.* Zhu, 2009:pp.296f.; Chen, 2000:pp.870-875) occur earlier is not inconsistent with this finding and simply suggests that the complete crystallisation and formalisation of 'washing' as a significant domain of sociocultural practices was a long-term process; (ii) explicitly allocated 'washing' vessels balance tradition and innovation: they make use of established types/shapes of bronze vessels, *i.e. ding*-tripods and *pan*-basins, yet bear inscriptions highlighting their function as implements of personal hygiene as well as their social significance; (iii) the inscriptions can be interpreted as the harbingers of a differentiated 'washing terminology' that became

---

23 I am grateful to Maria Khayutina for pointing out that the transliteration and interpretation of teng 滕 as wan 浣 in the *Digital Archives of Bronze Images and Inscriptions* might be incorrect. The character could be read as ying 媵 (dowry) – i.e. in the sense that the woman commissioned the *pan*-bassin for herself on the occasion of her marriage. However, teng 滕 also has an individual meaning; according to the *Shuowen* 'Teng also means gushing water' (水超涌也). If we accept this reading, the vessel could also be addressed as 'basin to collect gushing (i.e. poured?) water' and thus as referencing washing practices.

24 Note that the familial relations between Lord Shen of Cai and the woman addressed as 'Great Elder Ji' are still subject to discussion; some scholars argue that she was his elder sister, while others point out that a vessel as dowry would only have been given to a daughter. For a summary of different viewpoints see Cheng & Wan, 2017; for more information about a grave undoubtedly holding the remains of a daughter of Lord Shen of Cai, see Qiao & Li, 2009.

'textualized' only in later times; and (iv) designated 'washing' vessels were associated with members of the elite and firmly rooted in the male sphere: two third of the examined bronze vessels were commissioned by men and reserved for being used by them.

The archaeological evidence further suggests that specifically designated 'washing' vessels were not prevalent throughout the entirety of the Zhou realm but mainly occurred in a specific area (cf. fig. 1); this distribution is consistent with the territory gradually incorporated by the state of Chu 楚. In order to better understand the significance of these objects, it is therefore necessary to place them into the context of the rise of Chu and its changing relationship with the Zhou.

The 6/5th century BCE marks the transition from the late Springs-and-Autumns period to the subsequent Warring States period and therefore a time in which the ruling house of the Zhou had become increasingly insignificant (cf. *e.g.* Major & Cook, 2017:p.99). Already during the preceding centuries, the Zhou kings had struggled to maintain a hold over the empire they had created by setting up 'satellite states' in the more remote areas shortly after their accession to power. These were governed by family members or loyal followers (cf. Li, 2013:p.131; Mayor & Cook, 2017:pp.101-103) – all of them officially enfeoffed by the Zhou kings and thus expected to pledge their allegiance, for instance during regular visits to the capital (Mayor & Cook, 2017:p.132).[25] From around the 8th century BCE onwards, these ties gradually loosened and the empire became fragmented into a plethora of different states, many of them exerting political and military power that was on par with the Zhou kings.[26] In their ambitious drive for more authority and a larger territory these local rulers also used the force of arms, eventually leading to the so-called Warring States period.

Chu, with whom the Zhou held mostly amicable relations until the end of the 9th century BCE,[27] eventually became one of these quickly growing states[28] (cf. Li, 2013:p.120; Major & Cook, 2017:p.107). With the decline of the royal house, Chu seized the opportunity and started expanding along the southern fringe of the Zhou realm. The state reached its largest extent around the 6th century BCE and remained influential for more than three centuries before finally being conquered by the armies from the state of Qin 秦 in 223 BCE.

Although later written sources characterise Chu as a state hesitant to embrace Confucian norms, even denouncing it as 'alternative, slightly barbarous' (Cook & Blakeley, 1999:p.1), archaeological finds show that the Chu elites clearly subscribed to standards and values set in the Zhou sphere. They used ritual bronzes, carried out ancestral rituals and adhered to sumptuary rules. Until the mid-Chunqiu period even, the material culture of Chu was largely congruent with that of the Central Plains: grave construction followed the

25 This narrative as well as the identification of the Western Zhou state and its government with 'feudalism' have been criticized by e.g. Li, 2003 and Khayutina, 2010. Mayor & Cook (2017:pp.119-120) provide a brief summary of key issues.

26 Li (2013:p.132) refers to this phenomenon as the 'centrifugal' effect of local rulers enjoying a high degree of administrative and political autonomy and reinforcing their hold on their regional state – a phenomenon that is inextricably tied to a general relaxation of genealogical ties with the Zhou royal court (see esp. Li, 2006:pp.110-121).

27 According to Li (2013: box 6.1., p.118), three pieces of Zhou oracle bones speak of the external state of Chu and two of them report the visit of the Chu leader to the Zhou centre.

28 Note that the term Chu originally designated an area outside of the Zhou sphere, inhabited by various Yi tribes (Cook & Blakeley, 1999:pp.2f.; Blakeley, 1999). It was not until the Eastern Zhou Period that Chu was 'incorporated into the multi-state system of the Heartland Region' (Cook & Major, 2017:p.107) For an overview of many aspects of Chu culture, see the contributions in Cook & Blakeley, 1999.

Zhou-tradition and large numbers of bronze vessels were used in the tomb furnishings (So, 1999:p.34ff). Thereafter, regional traits – partly derived from the cultures and groups with which Chu interacted – became gradually apparent, making it difficult to define a coherent set of material culture in the area of Chu influence[29]. The weakening ties to the royal house of the Zhou (cf. Li, 2006:pp.110-121) and Chu's growing political and territorial power surely helped to install the self-confidence to turn away from the old ways, searching for more 'individualised' means of expression. Early indications of this new-found freedom might be seen in bronze vessels which, although adhering to well-established morphological schemes, were decorated with intricate ornaments and partly cast using a different technique, the lost-wax method (So, 1999:p.35).[30] Over time, however, these ritual bronzes lost their importance as grave goods and were gradually replaced by regional materials (cf. Xu, 1999:esp. pp.26-32) – in burials of the 4th century BCE such as Mashan (Hubeisheng Jingzhou diqu bowuguan, 1985) or Baoshan tomb 2 (Hubeisheng Jingsha tielu kaogudui, 1991) for instance, lacquer vessels and silks outnumber bronzes by far.

If we see the emergence of washing bronzes in this historical nexus, we can first and foremost interpret them as material expression of growing independence and confidence of the Chu elites[31] in a time of heightened socio-political volatility and re-configuration. The objects elegantly bridged the gap between tradition and innovation, underlining both their owners' awareness of the ancestral rites as well as their power to re-interpret these. Why, though, one may ask, the emphasis on personal cleanliness?

It is notable that the emergence of bronze washing vessels in Chu not only coincides with the reclining importance of the Zhou kings, it also falls into a time in which the quarrels between the states of the Zhou realm was nearing a peak. In this climate of

29 The variability of the material culture is aptly subsumed in Constance Cook's statement '(...) a culture that might be defined as Chu at one point in time would not necessarily fit the Chu of another point in time. One is forced to conclude, in fact, that there was no single real Chu.' (Cook & Blakeley, 1999:p.5).

30 In this context, see also Lothar von Falkenhausen (2006: pp.340-343), who points out that Xiasi tombs M1 and M2 contained two distinct assemblages of vessels, one "ordinary" and the other "special". While the former comprised vessels that seem to have been more or less standard in Zhanguo-period elite tombs of the Chu sphere and occurred in all bronze-yielding tombs at Xiasi, the latter is specific to tombs M1 and M2. The vessels forming part of this special assemblage are especially striking due to their size and lavish ornamentation as well as the technical sophistication with which they were executed. They are also "archaic in their shapes and in some of their decorative motifs (though not in their stylistic or technical execution!), and ... pointedly and no doubt deliberately refer to specific types of bronzes promulgated by the Late Western Zhou Ritual Reform." (Falkenhausen, 2006: p.342). In burials such as, e.g., Baoshan tomb 2, (Jingmen, Hubei) and Changtaiguan tomb 1 (Xinyang, Henan), some vessels of the "special assemblage" vessels are usually provided as ceramic mingqi whereas vessels belonging to the "ordinary assemblage" are functional. To von Falkenhausen (2006: p.350) this indicates that "to their owners and perhaps in general, "Ordinary Assemblage" vessels were of immediate practical utility in ritual, whereas the importance of the "Special Assemblage" may by then (or at this social level) have been mostly symbolic." The designated washing vessels from tombs M1 and M2 (M1:60, 72; M2:51-53, 55, 56; see appendix) belong to the ordinary assemblage, suggesting that they might have played a role in ritual activities.

31 Bearing in mind the diverse influences visible in Chu's material culture as well as the fact that protagonists mentioned in the inscriptions on washing vessels are addressed/address themselves as elites of Zeng, Deng etc., the prevalence of an overarching 'Chu' identity may appear unlikely. Research from psychology and the social sciences, however, highlights that instead of conceiving of identity as a single coherent attribute, identities are better envisioned as 'intersectional' (cf. Butler, 1990; for the intersectionality of 'body,' see e.g. Winker & Degele, 2010). Departing from a general critique of an overly 'homogenous' conception of culture, Welsch (2005:pp.326f.) similarly concludes that the category of 'identity' has always denoted a multi-stranded reference; cultural identities are built up within a web of often contradicting inter-linkages – it is these *plural* inter-linkages which define the position of an individual in its 'social field' (cf. Martin, 2003).

global competition, regional rulers sought to surpass each other on all possible vectors of society – political tactics and military struggles were accompanied by what we may call 'intellectual' and 'ideological warfare'. This created a situation in which the past could not legislate over the present anymore and in which political claims and aspirations could no longer be legitimised by referring to a line of succession or ancestral inheritances. Instead, the present became the focal point of society. The gradual shift in the connotation of bronze objects from explicit ritual devices of ancestral worship to material anchors of extended elite-networks illustrates this general re-configuration (see *supra*); it highlights that traditional bronze vessels had become an effective means of expressing social needs in the political present – a present that had become as contested as perhaps never before.

Numerous developments and innovations in the fields of technology, military strategy, administration, literature etc. were made possible in this climate. One of these novelties were activities in regards to water management. A response to an increase in population numbers and the growing need for enhanced agricultural productivity, on the one hand, they were also based on the insight that access and control over water(ways) could be a powerful political and military weapon[32] that enabled its users to better 'control' the landscape – both in the military and economic sense of the word – and to tighten territorial claims. In this spirit, the state of Chu carried out large-scale irrigation projects. Most of these undertakings are solely known through written sources[33], their archaeological detection being complicated by active alluvial processes (Zhuang, 2017:p.8). Nonetheless, traces of a several hundred-meter-long levee along the abandoned Yellow River channel in Northern Henan Province (Zhuang & Kidder, 2014) provide material evidence for the importance of 'domesticating water' at the time. Waterways and their strategic value may have also played a role in military decision-making, as is exemplified through the conquest of pre-existing waterways in the course of Chu's expansion[34].

The very practical task of 'cultivating' land tied into general questions about moral behaviour, 'good living' and the maintenance of social order through appropriate ways of governance that were raised by a new generation of thinkers. These scholars, some of them travelling savants, were actively patronised by local rulers or elites, encouraging intense debates and new ideas[35]. Political players sought to exploit some of this newly available knowledge to reinstall, legitimate and/or consolidate their power (*e.g.* Vankeerberghen, 2001). Aside from advancing classic technologies, in their quest for supremacy, they paid increasing attention to also furthering 'technologies of knowledge'.[36]

32 This is prominently subsumed in Needham, Wang & Lu (1971:p.265), who states: 'The feudal states also competed violently with one another for irrigation water, and in appropriating drained areas for growing crops.'

33 Apart from mentioned building activities, the emergence of specific designations for people concerned with water, e.g. inspectors of rivers and canals (*chuanheng* 川衡) provide further evidence for these developments. For more information and further literature, also on excavated seals with respective denominations, see Needham, Wang & Lu, 1971:p.267; Zhuang, 2017: table 5, note 83.

34 By conquering the states of Chen 陳 and Cai 蔡 in 479 and 447 BCE respectively, Chu gained access to the complex of artificial waterways that were to become the Canal of the Wild Geese (*Honggou* 鴻溝; Needham, Wang & Lu, 1971:pp.267-270; Bishop, 1997:pp.20f.).

35 For a novel publication on the intellectual dynamics of the Warring States period, which stresses common discourses rather than individual viewpoints of different schools of thought see Pines, 2009. For a classical approach on the topic see Fung, 1952; Fung, 1953.

36 See also Harper (1998:p.9), who suggests that 'Warring States natural philosophy' was originally the domain of practitioners of technical arts like calendrics, astrology and divination, whose views were most influential in the new view of nature.

During this time, early Chinese thinkers developed potent concepts such as 'Yin and Yang' 陰陽 and the 'Five states of matter' (*wuxing* 五行), which were eventually synthesized into what is later referred to as 'correlative cosmology' (*e.g.* Graham, 1986; Csikszentmihalyi, 2004; Kalinowski, 1991; Sun, 2000). This way of perceiving the world establishes a tight correspondence between individual, society and nature, thereby fostering the idea that shaping one of these domains influences the state of all others.[37] In this view, cosmic order is ensured through the harmonic adjustment of the various threads of reality, so that the manipulation of a single thread resonates with the others: the cultivation of nature may lead to the flourishing of society, the cultivation of society to the blossoming of the individual, and so forth. This conception opened up the possibility for advancing one's own cause by investing into coordinates of reality, which – until then – had received marginal attention.

In this particular milieu of thought, technologies may come into focus as 'techniques', defined by their *efficacy* to articulate change (*sensu* Mauss, 1950:p.371; cf. Warnier, 2009). This efficacy has at least two dimensions: on the one hand, techniques provide immediate material benefits, they for instance help to fertilise agricultural land or to facilitate military displacement; on the other hand, they can be expected to have collateral effects on the architecture of cosmos and society as a whole; technology thereby becomes a means to interfere with cosmic fate. Arguably, the surge of water technology and various technologies of knowledge in the time of interest is significant in this dual sense.

It is not inconceivable that in this particular socio-intellectual environment the human body was re-discovered as a coordinate of social reality, as a potent medium to articulate change on the political and cosmic levels. The documented rise of bronze objects tied to highly specific and thus standardised washing practices may then simply denote the growing significance of 'bodily techniques' in the concert of society-making. In a similar way as cultivating the land would contribute to the advancement of one's political cause (cf. Marks, 2012:pp.66-68), cultivating the body would strengthen one's own position in the greater cosmic game. The domestication of water and land, if you will, ties in with the 'domestication' of the human body.

These ideas are consistent with the growing prominence of the body in Warring States thought, which has been interpreted as the 'bodily turn' of ancient China (Brindley, 2010:pp.29f.). Scholarly work has shown that, from the 4th century BCE, the 'material virtue' discourse in *Ru* 儒 traditions emphasised the physical correlates of virtues and the ability to enact, through self-cultivation, change in the world and other people (cf. Csikszentmihalyi, 2004). At the same time, microcosm and macrocosm were theorised to 'resonate' and thereby shape the relationship between body, state and cosmos (Sivin, 1995). This growing significance of the body in navigating the world is also echoed in the surviving early medical manuscripts of the time.[38] These texts may be regarded as a first expression of an emerging

---

37 Although not consolidated in this way before the late 3rd century BCE, the early proponents mentioned above were present during the Warring States period) and their influence is clearly visible in Qin- and Han-dynasty texts, such as e.g. those found at Shuihudi (sealed ca. 217 BCE) and Mawangdui (sealed 168 BCE). For a brief overview of additional medical texts dating from the 3rd to the 1st century, see Harper, 1998: esp. pp.30-36.

38 These archaeological finds stem from Qin- (221-206 BCE) and Han-dynasty graves (206 BCE-9 CE) and are thus later in date than the washing bronzes. However, it is generally agreed that the texts draw upon a textual corpus that was circulating before that time (Harper, 1998:pp.19f.; Lo, 2007:pp.383f., n.4).

corpus of 'body technology' – what Unschuld (1980) and others have endeavoured to capture with the notion of 'medicine of systematic correspondences' (cf. Hsu, 2005:p.12). The proliferation of figurative decoration in Chu lacquer objects (cf. Fong, 1988-89; So, 1999: esp. p.40ff.) may thus be more than merely coincidental, perhaps similarly pointing to the growing importance of the human body in policing social and political realities.

Putting all of these bodily significances back into the outlined climate of global societal competition perhaps allows us to recognise the human body as a locus of political contest; the body may have become a means to exert worldly power and to display corporeal superiority. The emergence of novel assemblages of objects, practices and textual discourses addressing bodily issues may in this way signal the gradual 'ideologization' of the body – a process in which the 'social body' would have come more and more into view as a 'political body' (Lefebvre, 1991:p.40; Douglas, 1974)[39].

To drive this point home, we may consider the human body as a construction site in the larger 'civilising process' (*sensu* Elias, 1976) leading from the late Zhou to the establishment of the Han empire. Following Elias, this process can be re-cast as a coalescence of *sociogenesis* and *psychogenesis* – the incremental co-formation of social networks and a set of basic cognitive orientations. This 'civilising process', in which the rise and fall of Chu marks an important episode, presupposes the crystallisation of social practices foreshadowing a lasting structure of society. Based on the foregoing considerations, we may cautiously hypothesise that a reflexive attitude towards the human body, incentivising new practices to canvass it, was a crucial part of this process. Washing and cleaning but also other forms of bodily exercise, championed by various flourishing schools of thought, probably indicate that 'bodily techniques' conveyed an important aspect of the 'civilisatory' efforts and aspirations of the time. The very idea to 'cultivate' one's body in order to live a good and ethical life later popularised by Confucian doctrines may have its early roots here. The emergence of washing practices and other 'bodily techniques' around the 6/5th century BCE may thus anticipate what should later become a 'cultural technique' (Kost, 2017). Recalling Macho's seminal characterisation of 'cultural techniques' cited before, this pattern is not surprising at all.

Even though some of these reflections must remain preliminary we can nevertheless conclude that the broader sociohistorical context of the 6/5th century BCE offered a whole range of potentially connecting factors to explain the emergence of a spatiotemporally confined genre of washing vessel. In contrast to other available explanations, this account would emphasize the role of bodily practices – understood as efficacious 'technologies of the cosmos'- in the making of society and the constant tinkering with inherited materials to create possibilities for something new; it would also highlight the need to explain the rise of these objects by situating them within the relevant context of the social and political developments of their time.

An example for the importance of this approach is the study of Chen Zhaorong (2000) who also examined washing vessels from the same time frame. Interestingly, although departing from a slightly different corpus of source materials, this author outlines fairly similar factual trends (Chen, 2000: esp. pp.883-887). Chen (Chen, 2000:p.901) also concludes that washing vessels were largely commissioned by men; consistent with the present study, in about two-thirds of the cases they were ordered for personal use. The remaining

39 For the concept of the body as centralized, politicized space see also Lewis, 2006: chap. 1.

vessels were given as dowry, sometimes also as post-marriage gifts for woman (cf. Chen, 2000: table 884). Given this general convergence of results, it is even more surprising that Chen does not consider any of the contextual factors outlined before. Instead, she (Chen, 2000:p.901) interprets the respective washing vessels as an expression of male wealth and social status. Females, who played an important role in forging and maintaining relations between lineages 'would use them [*i.e.* the washing vessels] to control their looks, with the aim to strengthen the feelings of their husband's family through their beautiful appearance.' The objects thus also signify the importance of female beauty.

These explanations are rather generic and impose a stereotypical picture of 'maleness' and 'femaleness' onto a past context whose 'otherness' should not be downplayed easily. The universality of this account clearly counteracts the specificity of the sociocultural context in which the vessels are documented.

As I have tried to show in this paper, it seems to be highly significant that the respective washing devices are systematically associated with powerful males. It suggests that washing practices had an important social and perhaps even political function in the context of male display. This *appropriation* of practices of hygiene for the purpose of exerting and exhibiting social and political status, by demonstrating the power to influence the future course of society, can only be understood if we broaden the view for the particularities of the sociohistorical context in which the washing vessels make their appearance. Thus, the historical significance of washing and 'techniques of the body' cannot be discussed in general terms, let alone from the perspective of our own time and its prejudices. Rather, we must be prepared for the circumstance that there is likely much underrated 'difference' to be found in the past and that especially quotidian practices are highly sensitive to their social enfolding. The case of early 'washing' bronzes presented here illustrates this conclusion.

Bibliography

Bagley, R. W. (1987) *Shang Ritual Bronzes in the Arthur M. Sackler Collections*. Ancient Chinese Bronzes in the Arthur M. Sackler Collections Vol. 1. Washington, Arthur M. Sackler Foundation.

Bishop, K. (1997) *China's Imperial Way. Retracing an Historical Trade and Communications Route from Beijing to Hong Kong*. Hong Kong, Odyssey.

Blakeley, B. B. (1999) The Geography of Chu. In: Cook, C. A. & Major, J. S. (eds.) *Defining Chu. Image and Reality in Ancient China*. Honolulu, University of Hawai'i Press, pp. 9-20.

Boltz, W. G. (1993) Shuo wen chieh tzu 說文解字. In: Loewe, M. (ed.) *Early Chinese Texts. A Bibliographical Guide*. 2nd ed. Berkeley, Society for the Study of Early China, pp. 429-442.

Bottéro, F. & Harbsmeier, C. (2008) The *Shuowen Jiezi* Dictionary and the Human Sciences in China. *Asia Major*, 21 (1), 249-271.

Brindley, E. F. (2010) *Individualism in Early China. Human Agency and the Self in Thought and Politics*. Honolulu, University of Hawai'i Press.

Butler, J. (1990) *Gender Trouble. Feminism and the Subversion of Identity*. New York, Routledge.

Cao, Jinyan 曹锦炎 (1984) Shaoxing Potang chutu xuqi mingwen ji qi xiangguan wenti 绍兴坡塘出土徐器铭文及其相关问题. *Wenwu* 文物, 1, 27-29.

Cao, Jinyan 曹锦炎 (2011) Peng She tongqi mingwen bushi 彭射铜器铭文补释. *Wenwu* 文物, 6, 94-95.

Chen, Mengjia 陈梦家 (1956) Shouxian Cai hou mu tongqi寿县蔡侯墓铜器. *Kaogu xuebao* 考古学报, 2, 95-123.

Chen, Zhaorong 陳昭容 (2000) Cong gu wenzi cailiao tan gudai de guanxi yongju ji qi xiangguan wenti – zi Xichuan Xiasi Chunqiu Chu mu de qingtong shuiqi ziming shuo qi 從古文字材料談古代的盥洗用具及其相關問題 – 自淅川下寺春秋楚墓的青銅水器自名說起. *Zhongyang yanjiuyuan lishi yuyan yanjiusuo jikan* 中央研究院歷史語言研究所集刊, 71 (4), 857-932, 949-954.

Cheng, Pengwan 程鹏万 & Wan, Jiajun 万佳俊 (2017) Lun Da Meng Ji de siming 论大孟姬的私名. *Guji zhengli yanjiu xuekan* 古籍整理研究学刊, 3, 16-18.

Cook, C. A. (1997) Wealth and the Western Zhou. *Bulletin of the School of Oriental and African Studies*, 60 (2), 253-294.

Cook, C. A. & Blakeley, B. B. (1999) Introduction. In: Cook, C. A. & Major, J. S. (eds.) *Defining Chu. Image and Reality in Ancient China*. Honolulu, University of Hawai'i Press, pp. 1-5.

Cook, C. A. & Major, J. S. (eds.) (1999) *Defining Chu. Image and Reality in Ancient China*. Honolulu, University of Hawai'i Press.

Csikszentmihalyi, M. (2004) *Material Virtue. Ethics and the Body in Early China*. Sinica Leidensia 66. Leiden, Brill.

Dong Quansheng 董全生 & Li Changzhou 李长周 (2004) Nanyangshi Wuzicheng yi hao mu ji qi xiangguan wenti 南阳市物资城一号墓及其相关问题. *Zhongyuan wenwu* 中原文物, 2, 46-48.

Douglas, M. (1974) *Ritual, Tabu und Körpersymbolik. Sozialanthropologische Studien in Industriegesellschaft und Stammeskultur*. Frankfurt a. M., Fischer.

Elias, N. (1976) *Über den Prozeß der Zivilisation*. Frankfurt a. M., Suhrkamp.

von Falkenhausen, L. (1999) The Waning of the Bronze Age. Material Culture and Social Developments, 770-481 B.C. In: Loewe, M. & Shaughnessy, E. L. (eds.) *The Cambridge History of Ancient China. From the Origins of Civilization to 221 B.C.* Cambridge, Cambridge University Press, pp. 450-544.

von Falkenhausen, L. (2001) The Use and Significance of Ritual Bronzes in the Lingnan Region during the Eastern Zhou Period. *Journal of East Asian Archaeology*, 3 (1), 193-236.

von Falkenhausen, L. (2006) *Chinese Society in the Age of Confucius (1000-250 BC). The Archaeological Evidence*. Cotsen Institute of Archaeology. Ideas, Debates and Perspectives 2. Los Angeles, University of California Press.

Feng, Shi 冯时 (2009) Weifuren ding mingwen ji xiangguan wenti (化阝)夫人(女嚣)鼎铭文及相关问题. *Zhongyuan wenwu* 中原文物, 6, 64-66.

Fung, Yu-lan (1952) *A History of Chinese Philosophy. Volume 1. The Period of the Philosophers (from the beginnings to circa 100 B. C.)*. Translated by Derk Bodde. Princeton. Princeton University Press.

Fung, Yu-lan (1953) *A History of Chinese Philosophy. Volume 2. The Period of Classical Lerning (from the Second Century B. C. to the Twentieth Century A. D.)*. Translated by Derk Bodde. Princeton, Princeton University Press.

Geng, Chao 耿超 (2016) Gong Dian pan mingwen ji xiangguan wenti zai tan 公典盘铭文及相关问题再探. *Zhongguo guojia bowuguan guankan* 中国国家博物馆馆刊, 2016 (7), 139-144.

Gernet, J. (1998). *A History of Chinese Civilization*. 2nd ed. Cambridge, Cambridge University Press.

Graham, A. C. (1986) *Yin Yang and the Nature of Correlative Thinking*. Occasional Paper and Monograph Series. Institute of East Asian Philosophies 6. Singapore, Institute of East Asian Philosophies.

Harper, D. J. (1998) *Early Chinese Medical Literature. The Mawangdui Medical Transcripts*. London, Kegan Paul International.

Henansheng bowuguan 河南博物馆, Xichuanxian wenguanhui 淅川县文管会 & Nanyang diqu wenguanhui 南阳地区文管会 (1981) Henan Xichuanxian Xiasi yi hao mu fajue jianbao 河南淅川县下寺一号墓发掘简报. *Kaogu* 考古, 2, 119-127.

Henansheng wenwu yanjiusuo 河南省文物研究所, Henansheng Danjiang ku qu kaogu fajuedui 河南省丹江库区考古发掘队 & Xichuanxian bowuguan 淅川博物馆 (1991) *Xichuan Xiasi Chunqiu Chu mu* 淅川下寺春秋楚墓. 2 vols. Beijing 北京, Wenwu chubanshe 文物出版社.

Hubeisheng Jingsha tielu kaogudui 湖北省荊沙铁路考古队编 (1991) *Baoshan Chu mu* 包山楚墓. 2 vols. Beijing 北京, Wenwu chubanshe 文物出版社.

Hubeisheng Jingzhou diqu bowuguan 湖北省荆州地区博物馆 (1985) *Jiangling Mashan yi hao Chu mu* 江陵马山一号楚墓. Beijing 北京, Wenwu chubanshe 文物出版社.

Hubeisheng wenwu kaogu yanjiusuo 湖北省文物考古研究所 & Suizhoushi bowuguan 随州博物馆 (2014a) Hubei Suizhoushi Wenfengta Dong Zhou mudi 湖北随州文峰塔东周墓地. *Kaogu* 考古, 7, 18-34.

Hubeisheng wenwu kaogu yanjiusuo 湖北省文物考古研究所 & Suizhoushi bowuguan 随州博物馆 (2014b) Suizhou Wenfengta M1 (Zeng hou Yi mu), M2 fajue jianbao 随州文峰塔M1（曾侯乙墓）, M2发掘简报. *Jianghan kaogu* 江汉考古, 4, 3-51.

Hunansheng bowuguan 湖南省博物馆 (1963) Jieshao ji jian guancang Zhoudai tongqi 介绍几件馆藏周代铜器. *Kaogu* 考古, 12, 679-682, 10.

Hsu, E. (2005) Tactility and the Body in Early Chinese Medicine. *Science in Context*, 18 (1), 7-34.

Jingzhou bowuguan 荆州博物馆 & Zhongxiangshi bowuguan 钟祥市博物馆 (2009) Hubei Zhongxiang Huangtupo Dong Zhou Qin dai mu fajue baogao 湖北钟祥黄土坡东周秦代墓发掘报告. *Kaogu xuebao* 考古学报, 2, 247-294.

Kalinowski, M. (1991) *Cosmologie et divination dans la Chine ancienne. Le Compendium des cinq agents (Wuxing dayi, Vie siècle)*. Paris, École Francaise D'Extreme-Orient.

Kassung, C. & Macho, T. (2013) *Kulturtechniken der Synchronisation*. Kulturtechnik. München, Wilhelm Fink.

Khayutina, M. (2010) Royal Hospitality and Geopolitical Constitution of the Western Zhou Polity. *T'oung Pao*, 96 (1-3), 1-73.

Khayutina, M. (2014) Marital Alliances and Affinal Relatives (sheng 甥 and hungou 婚購) in the Society and Politics of Zhou China in the Light of Bronze Inscriptions. *Early China*, 37, 39-99.

Kost, C. (2017) Von Schmutzfinken, Saubermännern und Barbaren. Überlegungen zur Stellung der Körperpflege im China der Han-Dynastie. In: Müller, S. & Selbitschka, A. (eds.) *Über den Alltag hinaus. Festschrift für Thomas O. Höllmann zum 65. Geburtstag*. Wiesbaden, Harrasowitz, 141-159.

Lefebvre, H. (1991) *The Production of Space*. Translated by Donald Nicholson-Smith. 1st ed. in English Translation. Oxford, Blackwell.

Lewis, M. E. (2006) *The Construction of Space in Early China*. Albany, New York, State University of New York Press.

Li, Feng (2003) 'Feudalism' and Western Zhou China. A Criticism. *Harvard Journal of Asiatic Studies*, 63 (1), 115-144.

Li, Feng (2006) *Landscape and Power in Early China. The Crisis and Fall of the Western Zhou, 1045-771 BC*. Cambridge, Cambridge University Press.

Li, Feng (2013) *Early China. A Social and Cultural History*. Cambridge, Cambridge University Press.

Li, Ling 李零 (1981) 'Chu Shuzhisun Peng' jiujing shi shui? 楚叔之孙倗究竟是谁?. *Zhongyuan wenwu* 中原文物, 4, 38-39.

Li, Ling (1991) On the Typology of Chu Bronzes. Transl. von Falkenhausen, L. *Beiträge zur Allgemeinen und Vergleichenden Archäologie*, 11, 57-113.

Liu, Zenggui 劉增貴 (1998) Zhongguo gudai muyu lisu 中國古代沐浴禮俗. In: Zhongyang yanjiuyuan lishi yuyan yanjiusuo 中央研究院歷史語言研究所 (ed.) *'Qingjie' de lishi yantaohui*《清潔》的歷史研討會. (printed script of a presentation that until now has not been published officially) 11th-12th June.

Lo, V. (2007) Imagining Practice. Sense and Sensuality in Early Chinese Medical Illustration. In: Bray, F., Dorofeeva-Lichtmann, V. & Métailié, G. (eds.) *Graphics and Text in the Production of Technical Knowledge in China. The Warp and the Weft*. Leiden, Brill, pp. 383-425.

Ma, Sai 马赛 (2016) Cong shougongye zuofang kan Zhouyuan yizhi Xi Zhou wanqi de bianhua 从手工业作坊看周原遗址西周晚期的变化. *Zhongguo guojia bowuguan guankan* 中国国家博物馆馆刊, 3, 18-27.

Macho, T. (2003) Zeit und Zahl. Kalender und Zeitrechnung als Kulturtechniken. In: Krämer, S. & Bredekamp, H. (eds.) *Bild – Schrift – Zahl*. Kulturtechnik. München, Fink, pp. 179-192.

Major, J. S. & Cook, C. A. (2017) *Ancient China. A History*. Abingdon, Oxon, Routledge.

Marks, R. B. (2012) *China. Its Environment and History*. Lanham, Maryland, Rowman and Littlefield.

Martin, J. L. (2003) What Is Field Theory? *American Journal of Sociology*, 109 (1), 1-49.

Mauss, M. (1950) *Sociologie et anthropologie*. Paris, Presses Universitaires de France.

Moser, J. (2012) The Ethics of Immutable Things. Interpreting Lü Dalin's *Illustrated Investigations of Antiquity*. *Harvard Journal of Asiatic Studies*, 72 (2), 259-293.

Mou Yongkang 牟永抗 (1984) Shaoxing 306 hao Yue mu chuyi 绍兴306号越墓刍议. *Wenwu* 文物, 1, 12-13.

Nanjingshi bowuguan 南京市博物馆 & Liuhexian wenjiaoju 六合县文教局 (1991) Jiangsu Liuhe Chengqiao Dongzhou san hao mu 江苏六合程桥东周三号墓. *Dongnan wenhua* 东南文化, 1, 204-211.

Nanyangshi wenwu kaogu yanjiusuo 南阳市文物考古研究所 (2011) Henan Nanyang Chunqiu Chu Peng She mu fajue jianbao 河南南阳春秋楚彭射墓发掘简报. *Wenwu* 文物, 3, 4-31.

Needham, J., Wang, Ling & Lu, Gwei-Djen (1971) *Physics and Physical Technology. Part III: Civil Engineering and Nautics*. Science and Civilization in China Vol. 4. Cambridge, Cambridge University Press.

Okamura, Hidenori 岡村秀典 (2005) *Chugoku Kodai Oken to Saishi* 中国古代王権と祭祀. Tokyo 東京, Gakuseisha 学生舍.

Qiao, Baotong 乔保同 & Li, Changzhou 李长周 (2009) Nanyang faxian Cai hou Shen fu 南阳发现蔡侯申簠. *Zhongyuan wenwu* 中原文物, 2, 81-84.

Qiu, Xigui (2000) *Chinese Writing*. Translated by Gilbert L. Mattos & Jerry Norman. Early China Special Monograph Series 4. Berkeley, The Society for the Study of Early China and the Institute of East Asian Studies, University of California, Berkeley.

Pines, Y. (2009) *Envisioning Eternal Empire. Chinese Political Thought of the Warring States Era*. Honolulu, University of Hawai'i Press.

Rawson, J. (1989) Chu Influences on the Development of Han bronze vessels. *Arts Asiatiques*, 44, 84-99.

Rawson, J. (1990) *Western Zhou Ritual Bronzes from the Arthur M. Sackler Collections*. Ancient Chinese Bronzes from the Arthur M. Sackler Collections 2. Cambridge, Massachusetts, Harvard University Press.

Rawson, J. (1999) Western Zhou Archaeology. In: Loewe, M. & Shaughnessy, E. L. (eds.) *The Cambridge History of Ancient China*. Cambridge, Cambridge University Press, pp. 352-449.

Ren, Xianghong 任相宏 (1998) Shandong Changqingxian Xianrentai Zhoudai mudi ji xiangguan wenti chutan 山东长清县仙人台周代墓地及相关问题初探. *Kaogu* 考古, 9, 26-35.

Shaanxisheng kaogu yanjiuyuan 山西省考古研究院 (2010) *Zhouyuan. 2002 niandu Qijia zhijue zuofang he Licun yizhi kaogu fajue baogao.* 周原. 2002 年度齐家制玦作坊和礼村遗址考古发掘报告. 2 vols. Beijing 北京, Shehui kexue chubanshe 社会科学出版社.

Shaughnessy, E. L. (1983-1985) The Date of the "Duo You Ding" and its Significance. *Early China*, 9/10, 55-69.

Shaughnessy, E. L. (1991) *Sources of Western Zhou History. Inscribed Bronze Vessels*. Berkeley, University of California Press.

Shaughnessy, E. L. (1999) Western Zhou History. In: Loewe, M. & Shaughnessy, E. L. (eds.) *The Cambridge History of Ancient China. From the Origins of Civilization to 221 B.C.* Cambridge, Cambridge University Press, pp. 292-351.

Shaughnessy, E. L. (2009) Lunar-Aspect Terms and the Calendar of China's Western Zhou Period. In: Wang-Riese, X. & Höllmann, T. O. (eds.) *Time and Ritual in Early China*. Asiatische Forschungen 153. Wiesbaden, Harrassowitz, pp. 15-32.

Sivin, N. (1995) State, Cosmos, and Body in the Last Three Centuries B. C. *Harvard Journal of Asiatic Studies*, 55 (1), 5-37.

So, J. F. (1995) *Eastern Zhou Ritual Bronzes from the Arthur M. Sackler Collections*. Ancient Chinese Bronzes from the Arthur M. Sackler Collections 3. New York, Arthur M. Sackler Foundation.

So, J. F. (1999) Chu Art. Link between the Old and the New. In: Cook, C. A. & Major, J. S. (eds.) *Defining Chu. Image and Reality in Ancient China*. Honolulu, University of Hawai'i Press, pp. 33-47.

Sun, Xiaochun (2000) Crossing the Boundaries between Heaven and Man. Astronomy in Ancient China. In: Selin, H. (ed.) *Astronomy Across Cultures. The History of Non-Western Astronomy*. Science across Cultures. The History of Non-Western Science 1. Dordrecht, Springer, pp. 423-454.

Sun, Yan (2010) Inscribed Bronzes in Early Western Zhou Tombs: Funerary Gifts, Gifting and Social Network. In: 二十年來新見古代中國青銅器國際學術研討會. International

Conference 5th – 10th November 2010, Chicago. [Online] Available from: http://cccp.uchicago.edu/archive/2010FanBronzesConference/2010FanBronzesConference_SunYan.pdf [Accessed 24th April 2018].

Unschuld, P. U. (1980) *Medizin in China. Eine Ideengeschichte*. München, Beck.

Vankeerberghen, G. (2001) *The Huainanzi and Liu An's Claim to Moral Authority*. Albany, New York, State University of New York Press.

Wang, Changfeng 王长丰 & Hao, Benxing 郝本性 (2009) Henan xin chu 'Wei furen ding' mingwen jinian kao 河南新出"(阝化)夫人(女嚣)鼎"铭文纪年考. *Zhongyuan wenwu* 中原文物, 3, 69-75.

Wang, Changfeng 王长丰 & Qiao, Baotong 乔保同 (2009) Henan Nanyang Xujialing M11 xin chutu Wei furen ding 河南南阳徐家岭M11新出(阝化)夫人(女嚣)鼎. *Zhongyuan wenwu* 中原文物, 3, 10-11.

Warnier, J.-P. (2009) Technology as Efficacious Action on Objects… and Subjects. *Journal of Material Culture*, 14 (4), 459-470.

Welsch, W. (2005) Auf dem Weg zu transkulturellen Gesellschaften. In: Allolio-Näcke, L., Kalscheuer, B. & Manzeschke, A. (eds.) *Differenzen anders denken. Bausteine zu einer Kulturtheorie der Transdifferenz*. Frankfurt a. M., Campus Verlag, 314-341.

Winker, G. & Degele, N. (2010) *Intersektionalität. Zur Analyse sozialer Ungleichheiten*. Sozialtheorie Intro. 2nd ed. Bielefeld, Transcript.

Xu, Shaohua (1999). Chu Culture. An Archaeological Overview. In: Cook, C. A. & Major, J. S. (eds.) *Defining Chu. Image and Reality in Ancient China*. Honolulu, University of Hawai'i Press, pp. 21-32.

Yin, Wei 殷伟 & Ren, Mei 任玫 (2003) *Zhongguo muyu wenhua* 中國沐浴文化. Kunming 昆明, Yunnan Renmin Press 云南人民出版社.

Yuan, Yanling 袁艳玲 (2009) Zhoudai qingtong liqi de shengchan yu liudong 周代青铜礼器的生产与流动. *Kaogu* 考古, 10, 68-77.

Zaozhuangshi bowuguan 枣庄市博物馆, Zaozhuangshi wenwu guanli weiyuanhui bangongshi 枣庄市文物管理委员会办公室 & Zaozhuangshi Yichengqu wenguan xinju 枣庄市峄城区文广新局 (2014a) Shandong Zaozhuang Xulou Dong Zhou mu fajue jianbao 山东枣庄徐楼东周墓发掘简报. *Wenwu* 文物, 1, 4-27.

Zaozhuangshi bowuguan 枣庄市博物馆, Zaozhuangshi wenwu guanli weiyuanhui bangongshi 枣庄市文物管理委员会办公室 & Zaozhuangshi Yichengqu wenguan xinju 枣庄市峄城区文广新局 (2014b) Zaozhuang Xulou Dong Zhou mu fajue baogao 枣庄徐楼东周墓葬发掘报告. *Haidai kaogu* 海带考古, 7, 59-127.

Zeng, Lingbin 曾令斌, Huang, Yuhong 黄玉洪, Xiong, Yan 熊燕, Ai, Lingli 艾玲莉, Yu, Le 余乐, Yang, Hong 杨红, Gao, Jun 高隽, Tan, Zhuqing 谭竹青, Huang, Fengchun 黄凤春, Guo, Changjiang 郭长江, Zhang, Chengming 张成明 & Xiang, Zhang 项章 (2012) Hubei Suizhou Yidigang Zeng Gongzi Quji mu fajue jianbao 湖北随州义地岗曾公子去疾墓发掘简报. Jianghan kaogu 江汉考古, 3, 3-26.

Zhang, Chunlong 张春龙, Hu, Tienan 胡铁南 & Xiang, Kaiwang 向开旺 (2004) Hunan chutu de liang jian dong Zhou tongqi mingwen kaoshi 湖南出土的两件东周铜器铭文考释. *Zhongguo lishi wenwu* 中国历史文物, 5, 33-35.

Zhang, Juncheng 张俊成 (2014) Gong Dian pan ming bu shi 公典盘补释. *Kaogu yu wenwu* 考古与文物, 3, 105-108.

Zhong, Baisheng 鍾柏生, Chen, Zhaorong 陳昭容, Huang, Mingchong 黃銘崇 & Yuan, Guohua 袁國華 (eds.) (2006) *Xin shou Yin Zhou qingtongqi mingwen ji qiyig huibian* 新收殷周青銅器銘文暨器影彙編. Taipeh 台北, Yiwen Press 藝文印書館.

Zhou, Fazeng 周发增, Chen, Longtao 陈隆涛 & Qi, Jixiang 齐吉祥 (eds.) (1998) *Zhongguo gudai zhengzhi zhidushi cidian* 中国古代政治制度史辞典. Beijing 北京, Shoudu Shifan daxue chubanshe 首都师范大学出版社.

Zhu, Fenghan (2009) *Zhongguo qingtongqi zonglun* 中国青铜器总论. 3 vols. Shanghai 上海, Guji chubanshe 古籍出版社.

Zhuang, Yijie (2017) State and Irrigation. Archaeological and Textual Evidence of Water Management in Late Bronze Age China. *Wiley Interdisciplinary Reviews. Water*, 4 (4), 1-15.

Zhuang, Yijie & Kidder, T. R. (2014) Archaeology of the Anthropocene in the Yellow River Region, China, 8000-2000 cal. BP. *Holocene*, 24 (11), 1602-1623.

Appendix: Sites, Objects and further Key Data considered in the Text

Site name (Province)	Grave no.	Object no.	Vessel type	Signature character	Date of tomb	Tomb owner	Content of inscription	No. on map	Literature (selection)
Shouxian (Anhui)	/	21	*fou*	*guan* 盥	late SA (after 491 BCE)	♂	Ownership; commissioned by father for his eldest daughter (?)	1	Chen, 1956
		22	*fou*				Ownership; commissioned by man for himself		
		27	*yi*						
Xianrentai (Shandong)	M5	46	*pan*	*guan* 盥	late SA (before 560 BCE)	♀	Ownership; wish for long-term use of vessel; commissioned by husband for his wife	2	Ren, 1998; Zhang, 2014; Geng, 2016
Chengqiao Middle School (Jiangsu)	M3	6	*yi*	*guan* 盥	Late SA	♂	Ownership; commissioned by man for himself	3	Nanjingshi bowuguan & Liuhexian wenjiaju, 1991
Wancheng (Henan)	M38	53	*yi*	*mei* 沬	Late SA	♂	Ownership; commissioned by man for himself	4	Nanyangshi wenwu kaogu yanjiusuo, 2011
		58	*ding*	*tang* 湯					
Wuzicheng (Henan)	M1	?	*ding*	*tang* 湯	late SA (after 530 BCE)	♂	Ownership; wish for long-term use of vessel; commissioned by man for himself	5	Dong & Li, 2004
Xiasi, cluster C (Henan)	M1	60	*fou*	*yu* 浴	late SA (570-521 BCE)	♀	Ownership; date; wish for long-term use of vessel; commissioned by woman for herself	6	Henansheng wenwu yanjiusuo, Henansheng Danjiang kuqu kaogu fajuedui & Xichuanxian bowuguan, 1991
		72	*fou*	*yu* 浴					
	M2	51	*fou*	*yu* 浴	late SA (after 574 BCE)	♂	Ownership; commissioned by man for himself		
		52	*pan*	*guan* 盥					
		53	*yi*						
		55	*fou*	*yu* 浴					
		56	*ding*	*tang* 湯					
	M3	1	*pan*	*guan* 盥	late SA (570-521 BCE)	♀	Ownership; date; wish for long-term use of vessel; commissioned by father for his daughter		
		4	*ding*	*yu* 浴			Ownership; wish for long-term use of vessel; commissioned by man for himself		
Xujialing (Henan)	HXXM11	11	*ding*	*mu* 沐	early WS	♀	Ownership; date; wish for long-term use of vessel; commissioned by woman for herself	7	Wang & Hao, 2009; Wang & Qiao, 2009; Feng, 2009
Shizishan (Shanxi)	M306	採3	*ding*	*tang* 湯	early WS	♂	Ownership; date; wish for long-term use of vessel; commissioned by man for himself	8	Yan, 1984
Huangtupo (Hubei)	M3	8	*pan*	*guan* 盥	late SA	?	Ownership; date; wish for long-term use of vessel; commissioned by father for his daughter	9	Jingzhou bowuguan & Zhongxiangshi bowuguan, 2009
Wenfengta (Hubei)	M18	2	*fou*	*mu* 沐		♂	Ownership; mention of ritual usage; commissioned by man for himself	10	Hubeisheng wenwu kaogu yanjiusuo & Suizhoushi bowuguan, 2014a; b
	M33	30	*pan*	*huan* 浣	mid- to late WS	♀	Ownership; date; wish for long-term use of vessel; commissioned by woman for herself		
	M61	11	*fou*	*mu* 沐		♂	Ownership; commissioned by man for himself		
Gaoquanshan (Hunan)	M1	4	*pan*	*guan* 盥	mid-SA	♂	Ownership; date; wish for long-term use of vessel; commissioned by man for himself	11	Zhang, Hu & Xiang, 2004
Yangjiashan (Hunan)	?	?	*yi*	*huan* 浣	SA	♂?	Ownership; wish for long-term use of vessel; mention of ritual usage; commissioned by man for himself	12	Hunansheng bowuguan, 1963
Yidigang (Hubei)	M6	4	*dou*	*mu* 沐	Late SA	♂	Ownership; commissioned by man for himself	13	Zeng et al., 2012

Why is Death defiling? Considering death-related Pollution and Cleaning in Central Asia

Jeanine Dağyeli

"*We start* [our work] *by saying, Oh venerable Father Ali, king of mankind (Ya hazrat Ali shohi mardon baba) and we also finish thus. We start at the head and we come out at the feet. I take the jug* [with water for the ablution] *into my right hand and pour water into the left. I wash my hands three times. Then I start the ablution with the left hand. I take a piece of cotton and wipe over a part of the body. After each stroke we throw the cotton away and take a new ball. We do not touch the corpse with our hands!*"[1]

Cleaning and preparing a corpse for burial belongs to an almost universal set of funerary practices.[2] These take a different hue everywhere but there is a common, basic assumption that dying, death and burial (sometimes broken down into primary and secondary burial) are phases in a liminal period that must be set apart from normal, mundane time. Like other transitory rites, mourning and burial practices accompany the passage from one status to a fundamentally different one, from life to death. As first elaborated by Arnold van Gennep (1909) and later Victor Turner (1967), the transitional, liminal state is a necessary intermediate phase in between these two statuses. Writing a few years before van Gennep's *Les Rites de Passage* (1909), Robert Hertz, albeit from a different angle, came to similar conclusions concerning this liminal or, as he called it, transition period. In his essay 'A Contribution to the Study of the Collective Representation of Death', published in 1907, Hertz distinguished conceptualisations and practices engendered by death

1 Interview with corpse washer Jumagül apa in the historic centre of Khiva, July 2007. Her and other personal names have been altered. The Ali invoked in the quote is the son-in-law of the prophet Muhammad, fourth caliph to the Sunni and first Imam to the Shiites, who is regarded as the patron saint of corpse washing in Central Asia (see Snezarev, 2003:pp.123, 125-126; Dağyeli, 2011:pp.81-83).

2 The bulk of my fieldwork and archival research on death-related services and practices was carried out between 2009 and 2011 in the cities of Bukhara, Khiva, and Samarkand thanks to financial support from the Gerda-Henkel-Foundation. Additional material comes from the Kyrgyz, Uzbek and Tajik parts of the Ferghana Valley.

in: Bredenbröker, I., Hanzen, C. & Kotzur, F (eds) 2020: *Cleaning and Value. Interdisciplinary Investigations*, Sidestone Press (Leiden), pp. 155-166.

into those that pertain to the body of the deceased, the soul and to the survivors while acknowledging temporal and material aspects that conjoin the passage of the soul with the transformations of the body (Hertz, 2007:p.67). Cleaning the physical body is often correlated with purifying the deceased in a spiritual manner. Strangely enough, ritual aspects of cleaning the soul have received much more scholarly attention with regards to the Islamic world than the 'profane', physical cleaning of the corpse.[3]

In Muslim Central Asia, life and death are envisioned as an ongoing, mutual relationship between the living and their deceased who are regarded as ancestors. Both the bodily and the spiritual form of existence depend on the care and support from the respective other. For the living, this means the separation of the mourners from social and personal daily routines, festivities etc. and their gradual re-integration after the end of the mourning period, the stages of which are marked by commemorative ceremonies. For the deceased, it means a separation from life which they leave as a dead body.

After the burial, the deceased is eventually re-integrated with the newly achieved status as an ancestor spirit. The proper handling of burial and subsequent commemoration ceremonies plays a central role in the portrayal of the moral, responsible and religious self for many Central Asians (see Privratsky, 2001:p.96f.). But what exactly is this proper handling? What constitutes, re-establishes and maintains cleanliness and purity in the context of death? The answers can be quite diverse based on different interpretations of Islamic prescriptions, established local or localised practices, ethnic and social background, gender, age and other markers. There exist, however, some general practices and beliefs shared throughout Muslim Central Asia.

When a Muslim dies, their body has to be ritually washed according to canonical prescriptions. This does not only relate to outward cleanliness but also to ritual purity. The corpse has to be washed three times. Outright body beautification is not permitted by religion but natural scents like basil, rose water, clove or musk are often added to the wash water, especially for the second and third stages. How to clean the body in a proper way is often taught either in informal settings or at the mosque and may slightly differ locally. The service of washing the deceased is usually performed by same-sex family members or religious personnel and is perceived as meritorious. It can also serve as an expression of emotional closeness to the deceased person[4]. While this task also requires some training of the correct procedure and the prayers to be recited while washing, it is neither regarded as a profession nor a polluting activity.

Among some communities in Muslim Central Asia, however, there is an intense fear of pollution emanating from the dead body and the washing of the deceased is left to professional corpse washers[5]. The most common names for these washers are *ghassol* (an Arabic word meaning washer) or *murdashuy* (literally, washer of the deceased). While

3 This kind of cleaning may be symbolic, too. In a conversation, Ingeborg Baldauf told me about female corpse washers in northern Afghanistan who symbolically stroked the dead body with cotton without ever touching it. The body was, moreover, covered by sheets.

4 See Dağyeli, 2013:p.69. Since this article focuses on the professional washers, aspects of love, valediction and gratefulness that close ones may experience during the washing as part of the process of coping with the bereavement cannot be treated here. My interlocutors from among professional corpse washers all had a non-emotional approach to their work.

5 See Dağyeli, 2013; Dağyeli, 2015; Snesarev, 1973:pp.121f. Halevi (2007: chapter 2) discusses this attitude as a Zoroastrian-turned-Shiite conception but this explanation does not sufficiently answer why in Sunni Central Asia certain groups employ professional corpse washers.

the term *murdashuy* is ubiquitous in ethnographic literature, it is today often considered impolite in direct parlance. Rather, avoidance terms like *pokshuy* or *pokchi* (those who restore purity by washing, purifiers) are employed. Even if the washers ritually clean themselves afterwards, in the eyes of these communities washing corpses leaves a taint on the corpse washer which is passed on to the proceeding generations of the occupational family – an inherited pollution that sticks even to those who shift to other, more prestigious professions (see Dağyeli, 2013:p.53; Goffmann, 1997:p.73; McMurray, 2012:p.128; Snezarev, 2003:p.129). This view resulted in practices of avoidance on both sides, marriage rules and other norms aiming at limiting interaction between corpse washers and the rest of society to the absolutely necessary minimum. Historically, the corpse washers formed an endogamous professional group largely secluded from others (Snezarev, 2003:pp.123-125; Dağyeli, 2013; Dağyeli, 2018:pp.67-69, 72f.). This resulted in the corpse washers developing into a marginalised group glued together by external stigmatisation but little internal coherence. Professional corpse washers in Central Asia were of different ethnicities – usually the same ones as their immediate neighbours – and shared the same religious denomination and lifestyle. The only thing that largely excluded them from social interaction was their profession. Although the most obvious discrimination against professional corpse washers subsided during Soviet times, washers are still regarded with a mixture of fear, unease and awe because of their alleged inherent impurity as well as their reputation of having extraordinary powers to heal the sick and to avert imminent death. It is probably due to operating in the liminal spaces between life and death that corpse washers are accredited with this unique ability. The work of those who remain in this profession is partly monitored by the state: one part of their cash income is paid only after the corpse washer presents a receipt signed by the bereaved party, proving that everything has been completed properly.

Ethnicity and religious affiliation have become the single most significant criteria in attempting to explain the differences in death rituals practices and why people conceive of death and dying differently from neighbours, friends or family members. Many of my interlocutors explained the 'deviant' cleaning practices of their fellow compatriots, especially in regards whether to engage a professional corpse washer or not, as a matter of ethnic difference between Uzbek and Tajik, or Uzbek and Kyrgyz (Dağyeli, 2013:pp.60-63; Dağyeli, 2015:pp.542f.). In the cities of Bukhara and Samarkand for example – both perceived as 'Tajik' cities in an Uzbek surrounding – the general attitude prevails that death is polluting and that the washing of the deceased should be performed by a professional corpse washer. The local explanation – that these different approaches are the result of different ethnic backgrounds and practices – seems convincing at first. According to this explanation, Tajiks think that a corpse is polluting and thus employ professional corpse washers to carry out the religiously prescribed ablution before burial, whereas Uzbeks perform the task with the help of their extended family or social network. When including other places like the town of Khiva in Western Uzbekistan into these considerations, it quickly becomes obvious that categorisations according to ethnic groups are too narrow an explanation. Khiva, despite being thoroughly 'Turkic', has long been considered a stronghold of the 'pollution-faction' and, up until today, corpse washers practice their profession in town and adjacent rural communities. A more productive interpretation regarding the different views of the polluting powers of death and dead bodies is offered by Bel'kis Karmysheva, an eminent Soviet ethnographer (Karmysheva, 1976). She did not regard the relatively recent ethnic denotations of Central Asians pertinent but rather turned to the historical background for explanation[6]. According to Karmysheva, groups of people with either a non-tribal or so-called 'early' tribal background differed culturally, socially and in other ways from those with a 'late' tribal background in Central Asia[7]. While the former predominantly turned to professional corpse washers, the latter would not pass over their deceased to any non-kin for washing (Karmysheva, 1986:pp.165-167).

In this article I argue that the perception of pollution which some Central Asian communities feel emanates from the dead body is a response to the different liminal phases of death and the procedural separation of the living from the deceased after a death has occurred. As the famous dictum by Mary Douglas goes, 'dirt is essentially disorder. There is no such thing as absolute dirt: it exists in the eye of the beholder. ... Dirt offends against order. Cleaning is about reproducing the symbolic reproduction of order.' (1995:p.2). Corpse washing is simultaneously about physical and ritual cleaning, restoring order and transmitting pollution. Once it has manifested, pollution does not simply disappear but relocates itself. While the corpse washer cleans the deceased from potential stains, blood, saliva or other exudations and, even more importantly, restores a state of ritual purity to the body, the corpse washer inevitably become polluted by the substance of the dead and through contact with ritually impure matter[8]. This is, as McMurray states, not just a minor aspect of the work but a defining feature of this occupation (2012:pp.127f.). The pollution emanating from the dead body did not only rivet on the pre-Soviet corpse washers although they were the most conspicuous group. Pallbearers, gravediggers and other service providers also had a bad reputation although very little is known about their work and lives. The question remains why death is conceived of at all as causing impurity which then has to be removed; in other words: cleaned?

Pollution, Contagion and Avoidance

A first answer is rooted in Islamic canonical scriptures as well as in local beliefs about dying. According to popular conception, Azrail, the angel of death, comes to take a dying person. In this act of life-taking, the angel sheds the blood of the dying. In Islamic canonical texts, blood is – like other discharges of the human body – an impure substance. For anyone who comes into contact with them, it negates the ritual purity obtained through the morning ablution. Although the blood which the angel sheds remains invisible for the living, it is believed to be there and to pollute the corpse as well as material objects remaining in the room where death is taking place. According to some, this blood can even fill the whole room up to the ceiling (see Andreev, 1953:pp.195f.). The ability to avert death that the corpse washers are credited with manifests itself during a ritual that serves the purpose of distracting death from a client. This ritual necessarily demands a bloody sacrifice, usually of a cock. The human blood that would have been shed by Azrail needs to be substituted with the blood

6 Ethnicity remains a sensitive topic throughout Central Asia and too large an issue to tackle in the frame of this article. Present-day ethnicities were only fixed during early Soviet times and, even then, the subject of much discussion. See Baldauf, 1991; Baldauf, 1992; Schoeberlein-Engel, 1994; Sabol, 1995.

7 'Late' tribes refers to Turkic groups that came to Central Asia after the Mongol conquest in the early thirteenth century.

8 This shows an interesting parallel to folk medical healing practices whereby the disease has to be relocated to an animal or even an effigy that is then chased or taken away and thus secluded from contact with the community. See Baĭalieva, 1972; During & Khudoberdiev, 2007.

of another living being. The sacrifice is made in order to distract death's attention from humans and works along the same lines as surrogate offerings during healing rituals.

Mostly, bedclothes and clothes of the dying person are viewed as polluted. In some cases, all objects in the room or even in the house have to be ritually cleaned or thrown away[9]. According to findings of Snesarev from the Uzbek province of Khorezm, fields where a corpse was discovered were not sown for a period of five years for fear that the impurity of the dead body would be transferred onto the grain. In other places, the soil was cleaned by pouring water on it five times, removing the top soil or waiting until the sun had purified the piece of land. As for polluted material possessions, there were several ways to clean and thus restore monetary and use value to them. Preferably, they were to be exposed to sun-, moon- or starlight for at least a day and often longer. The light of celestial bodies, especially the sun, counted as potent purifier (Snesarev, 1973:p.121).

Water also has purifying properties when sprinkled on things. Fumigation with ispand, a Central Asian rue plant, is employed to clean rooms and objects. Other precautions are also taken. If death is imminent, foodstuff and other objects were and are still removed from the room or even the house where a dying person lies (see Kisliakov & Pisarchik, 1976:p.119; Bunn, 2008:p.17; Kalandarov & Shoinbekov, 2008:p.68). Some things cannot be cleaned. Food is a case in point. If people do not manage to remove food in time, it is often thrown away – which is otherwise absolutely rare – or it is fed to animals, usually to dogs in order not to waste it completely. Dogs are the only household animals which remain outside of the human food circle. They provide neither milk nor meat for human consumption and while the dung of cows and horses is often processed into fuel for cooking or used as fertiliser on the fields, dog excrement is not. Some other objects are given away or abandoned completely. Bed clothes and the deceased's clothes are often handed over to the corpse washer as part of the remuneration. In Khorezm and adjacent areas in Qaraqalpakistan, even the bier which only indirectly comes into contact with the deceased after the ritual cleaning is left next to the grave or abandoned. In other places in Uzbekistan, it is stored at the mosque or the graveyard and re-used for other funerals.

Maybe owing to the impediment of religious knowledge during the Soviet period, the explanation that the angel Azrail's shedding of blood was the reason why death is polluting, was voiced only occasionally during my research. Many stressed the allegedly modern hygiene discourse for not washing deceased people in the family over the more theological argument. Usto Shokir, a long-time interlocutor from Bukhara, connected the purity or pollution of the body with the spiritual purity of his city: 'The city of Bukhara has always been a stronghold of Islam and as such has remained extraordinarily pure. Now, when people die, you can never be sure whether they have died from a contagious disease or not. Therefore, out of reasons of hygiene and cleanliness, the Bukharans have special people who care for the washing of the dead' and, he added, for carrying the corpse to the graveyard. Where otherwise male relatives, friends and colleagues carry the bier in turns and conceive of this as religious merit, Bukharans often employ professional pallbearers who were also traditionally considered a stigmatised group[10]. Usto Shokir's

9 Dealing with bedclothes in this way stands in remarkable contrast to practices among other parts of the population where especially pillows of the deceased are seen as a medium to build up contact after death. See Dağyeli, 2014/2015:p.144.

10 According to Maria Louw (2007:p.70), ethnic Lūli, a low-prestige, traditionally peripatetic group have partly overtaken pall-bearing in Bukhara but this seems to be a recent development.

explanation implies one of the central factors regarding the importance of cleanliness in death rituals: To Central Asian Muslims, Islam is the religion of cleanliness and purity. This makes cleanliness valuable in itself; in filth, true faith cannot thrive. This general concern is amplified in settings where the dead body is considered polluted and polluting. Mary Douglas found this equation of cleanliness and 'good' *contra* pollution and 'bad' to be ubiquitous. '[I]n all societies the concept of dirtiness counters cleanliness on a moral level: while people tend to associate cleanliness with "goodness", dirtiness is largely associated with "badness".' (1995). Several scholars have pointed out that this ascription of badness does not only taint dirty activities but also the people who undertake them because of the negative connotation of dirt (see Selmi, 2012:p.113).

When watching the different stages of handling a dead body closely, it becomes obvious that the danger emanating from a corpse is not considered first and foremost a health issue, nor are the germs that might have caused the death. Ptomaine and unpleasant smells do not result in fear of the deceased. It is rather the belief that the substance of this person transfers upon surrounding objects, especially food, and thereby permeates everything surrounding a dying or deceased person. The belief that objects are perceptive to the negative influences which stem from a corpse was elaborated upon by Caroline Humphrey during her research in Mongolia. According to Humphrey, because of food's affectation of the perceptive faculties, it plays an important role in mediating between the relationship of the living and the dead (Humphrey, 2002). The intimate relation of a person's energy or spirit with objects, especially food, was also confirmed to me by a *Bibi Mullo* from Bukhara, a female religious specialist who assisted in death-related rituals. When describing a mourning ritual taking place forty days after death (*chihilum, qirq*) she said she would prepare a certain dish called *g'arma*, plain flat cakes made from flour and oil. Each guest should eat from them. After the meal, the plates on which the *g'arma* were served are rinsed without using soap. The wash water is collected and slowly poured into running water outside. If it is poured too fast, the guests are said to experience unpleasant feelings in their hands because some substance of each of them – their wishes on behalf of the deceased as well as their prayers – vest into the water (Dağyeli, 2015:p.559). Food to take home, distributed during mourning rituals, also absorbs, or literally 'is touched' by spiritual merit, acquired by participation and through prayer (Hardenberg, 2016: p.139). On the other hand, the deceased's substance also permeates things in the environment and has an ambivalent effect which people from nearby places might interpret in opposite ways. In the Kosonsoy district in the Ferghana Valley, the water used for washing the corpse, for instance, is disposed of at a far corner of the yard or in some other unfrequented place. If used to water fruit plants, they are believed to be in danger of withering (Ashirov, 2007:p.113). In the nearby Oltiariq district it is the opposite: people think this water should be poured under fruit trees (Karimova, 2013).

Like the body, the soul of a deceased person also has to be cleaned. According to religious beliefs, humans over the age of fourteen are fully responsible to God for their deeds. Since it is assumed that nobody remains without sins during their lifetime, several cleaning rituals for the soul are carried out during the burial and at commemoration gatherings. The first and foremost act is reading the Qur'an, praying and asking God for mercy on behalf of the deceased, in order to ease their burden in the grave. It is believed that the prayer is more effective if many people participate, especially in the graveyard prayer (*janoza*) which is carried out only by men during the burial ceremony. Thus, there is considerable social pressure on people to attend a funeral. This belief has been challenged by adherents of modernist or Salafi trends in Islam, maintaining that every individual is alone responsible for their deeds and that intercession is useless. Their rejection of traditional funerary practices is even more pronounced when it comes to crying, wailing and lamenting, especially if this is performed by professional lamenters. This habit is labelled outright un-Islamic. Many people, however, believe that certain forms of lamenting like praying is an effective intercession and therefore might even order lamenters prior to their death out of fear that only few people turn up for the burial (Dağyeli, 2014:p.141).

A ritual directly addressing the cleansing of the soul is called *davra*, the 'taking over of sins' (see Babaeva, 1985). It has been attested in ethnographic writings from many places of Central Asia into the twentieth century. Though rarely if at all practised today, it adds to an understanding of why corpse washers were and somehow still are considered polluted and stigmatised. During this ritual, somebody – often the corpse washer – declares himself willing to take over the deceased's sins to ease the burden this person may face in the afterlife. Using the corpse washer as a proxy to divert punishment for committed sins away from the deceased contradicts the above-mentioned purist Islamic reasoning regarding individual responsibility, which may be one of the reasons why the ritual falls more and more into oblivion. Shrinking numbers of professional corpse washers surely contribute to the abandonment of this ritual and a related one, the so-called 'paying off of debts' (*isqat*). *Isqat* has a similar function of alleviating debt, this time material debt which would still taint the deceased's soul if not returned[11]. The *isqat* could be practiced in different ways but essentially means distributing small amounts of money, matches, needles or other small things the deceased might have borrowed during their lifetime and not given back. Strangers and beggars were also included in this distribution of small items, bread and raisins. Additionally, the family of the deceased had to pay off real and more substantial debts the deceased might have accumulated. Maintaining good relations with the dead is important for a family's well-being. This is again done through food, preferably dough balls (*bog'irsoq*) fried in sizzling hot oil, however any dish that involves fuming oil is suitable because the souls of the deceased are said to be nourished by the fume. The food prepared in this oil is communally eaten later. Before the general availability of electric light the remains of oil were used for oil lamps, especially those used in commemoration rituals. Strengthened by the oil, the soul was able to set out for its weekly journey to Mecca and return frequently to its former house, bestowing blessing onto the family which was in turn cleaning their souls.

If the dead are not properly cared for either during burial or afterwards, they may inflict calamity upon the living. Since the power of the dead is deemed to be so great as to enable them to reach out for the living and take revenge for not being correctly handled, great responsibilities are incumbent on the corpse washers. In one instance, so I was told during fieldwork, a corpse was even taken out of the grave again and cremated, otherwise completely forbidden for Central Asian Muslims. After his initial death several people from his close family also died within a relatively short time and these tragic events were explained through possible misconduct in the handling of the corpse. The case was eventually brought before a local court because of the cremation but I was not able to

11 See Bellér-Hann (2008) for this ritual among the Uyghur; Hardenberg, 2016:p.129 for the Kyrgyz.

Figure 1. Receipt of a corpse washer (Photo: Jeanine Dagyeli).

trace it down to its close. Historically, disinterment was likewise practised in the rare and extraordinary circumstances when there was fear of the dead's revenge. When an unknown disease broke out in the small mountain settlement of Anzob (today Tajikistan) in autumn 1898 and roughly two thirds of the inhabitants passed away within a short period of time, rumour spread that the epidemic proliferated because its first victim – a woman who had washed a corpse in a neighbouring village (we are not told whether she was a professional corpse washer or not) – had not been buried correctly. Finally, even the local mullahs approved of opening the grave of the corpse washer who had apparently

Figure 2. Abandoned bier at the graveyard Mizdahan, Autonomous Republic of Qaraqalpaqstan (Photo: Jeanine Dagyeli).

brought in the disease in order to prove that she had been buried according to the rules of the shariᶜa and that the rumours were invalid (see Iskandarov, 2012:pp.434-435; *Turkestanskie Vedomosti*, 08.11.1898 (44)).

Conclusion

Can cleaning be conceptualised as an analytical tool? The answer to this question probably cannot be packed into a generally valid formula but rather has to be firmly rooted in the experiences, practices and world-views of those we work with. In the case outlined above, this means that death-related cleaning encompasses both the material world – the body, a room etc. – and the immaterial world, the soul. In all of these different cases, Tajik and Uzbek speakers use the word *pok* (clean, pure) to refer to both the physical removal of dirt, blood etc. and to the removal of invisible pollution, such as sin and debt. While cleaning is obviously performed in multiple ways as a service for the dead by the living, the dead reciprocally help clean the souls of the living by bestowing blessings and intercession with God if they are properly cared for. Cleaning makes and (re-)orders the relations between the living and the dead. With regards to the inanimate world, cleaning does the same thing. It restores the relations between people and things which have been disrupted by the pollution caused by death. Cleaning returns value to things and makes their return into the sphere of human interaction possible.

Most people among my interlocutors would agree that "dirty work", especially if human substances are involved, is something to be avoided. In their understanding, work that has physically and spiritually polluting faculties would classify as "dirty". Since this kind of labour also tends to come with little or no social esteem and limited income, "dirty work" and menial labour (called *qora ish*, literally "black work" in Uzbek) are the resort of those with few other options. These kinds of occupations are not voluntarily chosen by those with career aspirations and life choices (McMurray, 2012:p.128). Why then do corpse washers become corpse washers? Historically, children born into families of corpse washers did not have many professional alternatives but to enter into this profession themselves. Unlike the corpses they washed, the washers themselves could not rid themselves of their pollution. Even after the period of explicit discrimination, from the 1930s onwards, Soviet ethnographers like Gleb Snesarev repeatedly encountered discrimination of corpse washers during the first decades under Soviet occupation. Even into the 1950s people referred to corpse washers as 'third class people' and one shrine guardian objected to Snesarev noting down information on Islamic saints and corpse washers in the same notebook (Snesarev, 1973:p.127; Snezarev, 2003:p.125). Today a corpse washer, especially if not born into one of these traditional families, may feel some necessity to explain their professional choice (see Dağyeli, 2013:pp.68f.). These tasks became increasingly unattractive to most members of the traditional, not ethnically marked communities of corpse washers. As more prestigious career opportunities also appeared, corpse washing vacancies became available for other marginalised groups like the Lūlī, a peripatetic ethnic minority referred to as Central Asian gypsies in academic and popular literature. If this turns out to be a longer lasting and widening trend, we might indeed witness an ethnicization of those 'unclean' death-related services, but an ethnicization different from how it is largely perceived today.

Bibliography

Andreev, M. S. (1953) *Tadzhiki doliny Khuf*. Vol. 1. Stalinabad, Izd-vo Akademii nauk Tadzhikskoĭ SSR.

Ashirov, A. (2007) *O'zbek xalqining qadimiy e'tiqod va marosimlari*. Tashkent, Alisher Navoiy nomidagi O'zbekiston milliy kutubkhonasi nashriyoti.

Babaeva, N. S. (1985) Davra – obrīad vykupa grekhov umershego. In: Litvinskiĭ, B. A. (ed.) *Etnografiīa Tadzhikistana. Sbornik stateĭ*. Dushanbe, Donish, pp. 58-64.

Baĭalieva, T. D. (1972) *Doislamskie verovaniīa i ikh perezhitki u kirgizov*. Frunze, Ilim.

Baldauf, I. (1991) Some Thoughts on the Making of the Uzbek Nation. *Cahiers du monde russe et soviétique*, 32 (1), 79-95.

Baldauf, I. (1992) *"Kraevedenie" and Uzbek National Consciousness*. Papers on Inner Asia 20. Bloomington, Indiana, Indiana University, Research Institute for Inner Asian Studies.

Bellér-Hann, I. (2008) *Community Matters in Xinjiang 1880-1949. Towards a Historical Anthropology of the Uyghur*. China Studies 17. Leiden, Brill.

Bunn, S. (2008) Küt and Dzhut. Food and Famine in Central Asia. In: Gervers, M., Bulag, U. E. & Long, G. (eds.) *The Exploitation of the Landscape of Central and Inner Asia. Past, Present and Future*. Toronto Studies in Central and Inner Asia 9. Toronto, Asian Institute, University of Toronto.

Dağyeli, J. E. (2011) *"Gott liebt das Handwerk". Moral, Identität und religiöse Legitimierung in der mittelasiatischen Handwerks-risāla*. Iran-Turan 12. Wiesbaden, Reichert.

Dağyeli, J. E. (2013) The Past as a Burden. Washers of the Dead between Merit and Stigma. *Asien. The German Journal on Contemporary Asia*, 129, 51-72.

Dağyeli, J. E. (2014) Contested Mourning. Central Asian Funerary Practices in Local and Global Islam. *Anthropology of the Contemporary Middle East and Central Eurasia*, 2 (2), 139-162.

Dağyeli, J. E. (2015) How to Do a Burial Right: Negotiations of Identity, Religious Practice and the State. *Archiv Orientální,* 83 (3), 537-567.

Dağyeli, J. E. (2018) Whose Margins? Marginality, poverty and the moral geography of pre-Soviet Bukhara. In: Chappatte, A., Freitag, U. & Lafi, N. (eds.) *Understanding the City through its Margins. Pluridisciplinary Perspectives from Case Studies in Africa, Asia and the Middle East.* Routledge Studies in Modern History. London, Routledge, pp. 63-76.

Douglas, M. (1995) *Purity and Danger. An analysis of the concepts of pollution and taboo.* 9th ed. London, Routledge.

During, J. & Khudoberdiev, S. (2007) *La voix du chamane. Étude sur les baxshi tadjiks et ouzbeks.* Collection Centre-Asie. Paris, IFEAC – L'Harmattan.

Goffman, E. (1997) *The Goffman reader.* Malden, Blackwell.

Halevi, L. (2007) *Muhammad's Grave. Death Rites and the Making of Islamic Society.* New York, Columbia University Press.

Hardenberg, R. (2016) Three Dimensions of Liminality in the Context of Kyrgyz Death Rituals. In: Berger, P. & Kroesen, J. (eds.) *Ultimate Ambiguities. Investigating Death and Liminality.* New York, Berghahn, pp. 125-144.

Hertz, R. (2007) *Das Sakrale, die Sünde und der Tod.* Wissen und Studium. Sozialwissenschaften. Konstanz, UVK.

Humphrey, C. (2002) Rituals of death as a context for understanding personal property in socialist Mongolia. *Journal of the Royal Anthropological Institute,* 8 (1), 65-87.

Iskandarov, B. (2012) *Vostochnaīa Bukhara i pamir vo vtoroĭ polovine XIX. v.* 2nd ed. Dushanbe, Donish.

Kalandarov, T. S. & Shoinbekov, A. A. (2008) Some Historical Aspects of Funeral Rites among People of Western Pamir. *Anthropology of the Middle East,* 3 (1), 67-81.

Karimova, M. B. (2013) Participations of G'assols (Washers of the Dead) in Mourning Ceremonies (in the Example of Ferghana Valley). *Molodoĭ uchenyĭ,* 1, 289-291.

Karmysheva, B. K. (1976) *Ocherki etnicheskoĭ istorii īuzhnykh raĭonov Tadzhikistana i Uzbekistana.* Moscow, Nauka.

Karmysheva, B. K. (1986) Arkhaicheskaīa simvolika v pogrebal'no-pominal'noĭ obrīadnosti Uzbekov Fergany. In: Basilov, V. N. (ed.) *Drevnie obrīady verovanniīa i kul'ty narodov Sredneĭ Azii. Istoriko-etnograficheskiye ocherki.* Moscow, Nauka, pp. 139-181.

Kislīakov, N. A. & Pisarchik, A. K. (1976) *Tadzhiki Karategina i Darvaza.* Vol. 3. Dushanbe, Donish.

Louw, M. E. (2007) *Everyday Islam in Post-Soviet Central Asia.* Central Asian Studies Series. London, Routledge.

McMurray, R. (2012) Embracing Dirt in Nursing Matters. In: Simpson, R., Slutskaya, N., Lewis, P. & Höpfl, H. (eds.) *Dirty Work. Concepts and Identities.* Identity Studies in the social sciences. Basingstoke, Palgrave Macmillan, pp. 126-142.

Privratsky, B. G. (2001) *Muslim Turkistan. Kazak Religion and Collective Memory.* Richmond, Surrey, Curzon.

Sabol, S. (1995) The creation of Soviet Central Asia: the 1924 national delimitation. *Central Asian Survey,* 14 (2), 225-241.

Schoeberlein-Engel, J. S. (1994) *Identity in Central Asia. Construction and Contention in the Conceptions of "Özbek", "Tâjik", "Muslim", "Samarqandi" and other Groups.* Cambridge, Massachusetts, Harvard University.

Selmi, G. (2012) Dirty Talks and Gender Cleanliness: An Account of Identity Management Practices in Phone Sex Work. In: Simpson, R., Slutskaya, N., Lewis, P. & Höpfl, H. (eds.) *Dirty Work. Concepts and Identities*. Identity Studies in the social sciences. Basingstoke, Palgrave Macmillan, pp. 113-125.

Snesarev, G. P. (1973) *Pod nebom Khorezma*. Moscow, Mysl.

Snezarev, G. P. (2003) *Remnants of Pre-Islamic Beliefs and Rituals among the Khorezm Uzbeks*. Mittelasiatische Studien 5. Berlin, Reinhold Schletzer Verlag.

Turner, V. W. (1967) Betwixt and Between: The Liminal Period in Rites de Passage. In *The Forest of Symbols. Aspects of Ndembu Ritual*. Ithaca, New York, Cornell University Press, pp. 93-111.

Turner, V. W. (1969) *The Ritual Process. Structure and Anti-Structure*. Chicago, Aldine de Gruyter.

van Gennep, A. (1909) *Les rites de passage. Étude systématique des rites de la porte et du seuil, de l'hospitalité, de l'adoption, de la grossesse et de l'accouchement, de la naissance, de l'enfance, de la puberté, de l'initiation, de l'ordination, du couronnement, des fiançailles et du mariage, des funérailles, des saisons, etc.* Paris, Émile Nourry.

Your Clothes Should Be Clean! Your Head Should Be Washed! Body Cleaning and Social Inclusion in the Epic of Gilgamesh

Ainsley Hawthorn

The world's first writing system, cuneiform, originated in ancient Mesopotamia, a region bounded by the Tigris and Euphrates rivers in roughly the same area as modern-day Iraq. Initially used for recording economic transactions, as cuneiform became more robust it progressed from simple notations of numbers and nouns to a fully developed method for recording every syllable of the area's spoken language, Sumerian, and soon scribes were using cuneiform to compose a variety of other texts, including royal inscriptions, legal documents, personal letters, and literature.

Stories of the Mesopotamian hero Gilgamesh are among the earliest that were committed to writing. Three poetic fragments detailing his exploits survive from the Ur III period (ca. 2112-2004 BCE) and, by the Old Babylonian period (ca. 2003-1595 BCE), scribes-in-training were studying Sumerian poems about the hero and beginning to record Gilgamesh legends in a second language, Akkadian, which is a Semitic language related to modern-day Arabic and Hebrew (George, 2003:p.7-8). These Akkadian narratives incorporate some material from their Sumerian predecessors but also introduce many new elements that may derive from the oral tradition.

Whereas each of the Sumerian Gilgamesh poems was self-contained, relating one specific story about the hero, the Akkadian texts begin to form a cycle, weaving individual episodes and adventures into a longer narrative and addressing themes of nature and culture, love and friendship, life and death. In the late second millennium BCE, this integrated Gilgamesh narrative became fixed into a twelve-tablet text known to us as the Standard Babylonian Gilgamesh Epic and to the ancients as *ša naqba īmuru*, 'He who saw the deep'. This version of Gilgamesh's story was faithfully copied for a thousand years and was housed in libraries and tablet collections throughout Babylonia and Assyria.

While the epic has been the subject of a great deal of scholarly discussion over the past century and a half, little attention has been paid to how episodes of body cleaning in the story reinforce the narrative's thematic content, conform to its literary structure, and reveal Mesopotamian social norms. There are five major and three minor instances of body washing in the epic. The major passages, which describe the cleansing process in some detail

in: Bredenbröker, I., Hanzen, C. & Kotzur, F (eds) 2020: *Cleaning and Value. Interdisciplinary Investigations*, Sidestone Press (Leiden), pp. 167-180.

over several lines, feature three characters: Enkidu, Ninsun and Gilgamesh. In addition to these lengthier examples, three lines in the epic mention bathing more briefly. This paper will review these instances of body cleaning, analyse their practical features and literary functions, and ultimately assess how they develop the themes of the epic composition.

Instances of Body Cleaning

At the beginning of the narrative, Gilgamesh is a thoughtless young king of the city of Uruk who uses his prodigious strength, massive size and political authority to abuse his subjects. Hearing the anguished cries of the citizenry, the gods decide to create the superhuman Enkidu from clay as a rival to Gilgamesh, although their plan ultimately backfires when Gilgamesh and Enkidu become fast friends (Hawthorn, 2015). Following his creation, the gods deposit Enkidu in the wilderness outside of the city, where at first he contents himself to run naked with the animals. Allied with the wildlife, Enkidu causes trouble for local hunters by filling in their pit-traps and dismantling their snares. On the advice of Gilgamesh, the hunters resolve to civilize Enkidu in order to alienate him from his wild companions.

First, they bring the prostitute Shamhat to the water hole where Enkidu drinks. When she strips off her clothing before the wild man, he is unable to resist her charms, and the two make love for six days and seven nights. On the seventh day, when Enkidu attempts to return to his herd, the animals flee, and he finds himself unable to follow. Although Enkidu's physical powers are diminished, in their place he has a new, human capacity for reason.

What happens next is reconstructed from earlier versions of the epic dating to the Old Babylonian period. Shamhat dresses Enkidu in some of her clothing and leads him by the hand to a nearby shepherds' encampment. The shepherds give him bread and beer, processed foods and staples of civilization that Enkidu has never tasted during his time in the wilderness. Then:

> The barber treated his hairy body.[1]
> He anointed himself with oil and turned into a man.
> He put on clothing and became like a warrior.
> He took up his weapon and sparred with lions.
> The shepherds lay down at night.
> He slaughtered all the wolves, he scattered all the lions.
> The senior herdsmen slept.
> Enkidu was their watchman, a vigilant man.
>
> (OB Pennsylvania, ll. 106-118)[2]

After experiencing the fundamentally human activities of eating bread, drinking beer, bathing, and dressing, Enkidu transforms from the guardian of the wild animals to their scourge, slaying wolves and scaring away lions to defend the flocks while the shepherds sleep.

The second episode of body cleaning involves Gilgamesh's mother, the goddess Ninsun. Gilgamesh and Enkidu, now friends, have informed her of their plan to travel to

---

1 This line has alternately been translated 'He washed his hairy body with water' (see, for example, Foster 2001). George has rejected the reading 'water' based on close examination of the tablet, and I follow his interpretation 'barber' (2003:p.185).

2 For editions of the Babylonian Gilgamesh texts see George (2003).

the forest of cedars in Lebanon to fight the monstrous guardian of those woods, Humbaba. Dismayed, she resolves to take the matter up with the sun god Shamash and prepares herself to meet with him:

> Into the bath-house she went, seven times.
> [She purified] herself in water with tamarisk and soapwort.
> [She dressed in] a fine garment, seemly for her body,
> [...] ... seemly for her breast.
> [...] was set in place, and she donned her crown.
> [...] ... the harlots... the ground.
> She skipped [up] the staircase, she climbed onto the roof.
> She climbed onto the roof, she set up an incense offering before Shamash.
> She placed an offering before Shamash, she lifted her arms:
> "Why did you endow my son, Gilgamesh, with a restless heart?"
>
> (Tablet III, ll. 35-46)

In this passage, Ninsun performs an ablution – the ritual cleansing of the body for religious purposes – and her actions reflect real-life religious practice. The bathing of the king before undertaking ritual duties is attested as early as the reign of Gudea at the end of the third millennium BCE (Edzard 1997:p.80 E3/1.1.7.CylA Col. xviii ll.3-9). Ninsun's seven separate sessions in the bathhouse, in particular, recall the Assyrian ritual series *bīt rimki* or "House of Ablution", which prescribes an ablution rite for the king that is divided into seven "houses" (Laessøe 1955; George 2003:p.459). Notably, most of the prayer in the *bīt rimki* ritual is addressed to Shamash, the same god Ninsun is petitioning in the epic.

Gilgamesh is the third character whose bathing is described in detail in the text. He and Enkidu have just returned from the forest of cedars where they slew the monster Humbaba. The next tablet opens with the following passage:

> He washed his matted hair, he cleaned his equipment,
> He shook his locks down over his back.
> He cast off his soiled things, he dressed himself in his clean things.
> He wrapped himself in cloaks, tied with a sash.
> Gilgamesh put on his crown.
> The princess Ishtar coveted the beauty of Gilgamesh:
> "Come, Gilgamesh, you will be the bridegroom!
> "Give me, oh give me of your fruits.
> "You will be my husband, and I shall be your wife!"
>
> (Tablet VI, ll. 1-9)

Here, Gilgamesh's freshly cleaned body attracts the erotic interest of Ishtar, goddess of love and war, who propositions him with a phrase that inverts the language of the traditional marriage proposal, which under normal circumstances would be "you will be my wife, and I shall be your husband", as attested in some magical texts (Lackenbacher 1971:p.126 Col. II ll.13-15, p.153). The very fact that it is Gilgamesh who bathes and Ishtar who makes a sexual overture may be a reversal of gender roles, in that normally it would be the woman who bathes in preparation for a sexual encounter with a man (Leick 1994:p.125). Recalling the many former lovers who have suffered at Ishtar's hands, however, Gilgamesh rejects her advances.

The fourth major instance of body washing in the epic again features the hero Gilgamesh. Having crossed the tunnel of the sun and navigated the waters of death to reach the home of the legendary survivor of the great flood, Uta-napishti, Gilgamesh is at last turned away without the secret to eternal life. Before Gilgamesh leaves, however, Uta-napishti instructs his servant, Ur-shanabi, to treat the hero as follows:

> "The man whom you brought here,
> "Whose body is overgrown with matted hair,
> "The beauty of whose flesh has been marred by hides,
> "Take him, Ur-shanabi, get him to the washtub,
> "Let him wash his matted hair with water, clean as can be.
> "Let him cast off his hides, and let the sea carry them away.
> "Rinse his body clean.
> "His headband should be renewed.
> "He should be dressed in a robe that befits his status.
> "Until he arrives in his city,
> "Until he reaches (the end of) his journey,
> "Let the robe show no stain but stay brand new."
>
> (Tablet XI, ll. 250-61)

So cleansed, Gilgamesh returns to Uruk in the company of Ur-shanabi.

The fifth and final substantial episode of body cleaning appears in the last tablet of the epic, an epilogue telling a story that is thematically related to the main text but does not form part of the epic narrative. Enkidu has volunteered to descend to the Netherworld to retrieve a ball and bat that Gilgamesh has dropped. Gilgamesh warns Enkidu to be cautious and gives him a series of instructions on how to behave so as to prevent the spirits of the dead from recognizing him as an outsider. Enkidu, however, is reckless:

> [Enkidu] went down [to the Netherworld],
> But he paid no heed to the instructions of Gilgamesh.
> He dressed himself in [a spotless garment] -
> They identified him as a stranger.
> He anointed himself with sweet oil from a bowl –
> They gathered at the scent [of it].
>
> (Tablet XII, ll. 31-36)

As a result of Enkidu's indiscretion, he is seized by the Netherworld and is unable to return to the world of the living.

In addition to the previous examples, which devote multiple lines to illustrating the cleansing procedure, three passages in the epic mention body cleaning in brief. The first two instances concern hand washing: the goddess Aruru washes her hands before creating Enkidu from clay, and Enkidu and Gilgamesh wash their hands in the Euphrates after killing the Bull of Heaven.

Aruru washed her hands.
She pinched off clay; she threw it down upon the steppe.

(Tablet I, ll. 101-102)

They washed their hands in the Euphrates.
They clasped each other and walked away.

(Tablet VI, ll. 167-168)

The third short reference to body cleaning appears in tablet XI, near the end of Gilgamesh's adventure. As Gilgamesh begins his journey back to Uruk, Uta-napishti tells him the location of a magical plant of rejuvenation so that he does not have to return home from his quest empty handed. After retrieving the plant, Gilgamesh makes toward Uruk with Ur-shanabi. When they pitch camp for the night:

Gilgamesh found a pool whose waters were cool,
And he went down into it to bathe in the water.
A snake smelled the aroma of the plant.
[Silently], it came up and carried the plant away.
As it turned away, it shed a skin.

(Tablet XI, ll. 303-307)

While Gilgamesh is distracted, a snake sneaks up to the camp and makes off with the plant, shedding its skin on the way as a testament to the plant's rejuvenating powers. Gilgamesh mourns the loss of the plant but ultimately turns his attention to the enduring works that he has created in Uruk: the massive walls and well-developed urban landscape that represent an attainable source of immortality insofar as they establish his eternal reputation.

Features of Body Cleaning

Evidence from a wide variety of text genres as well as archaeological finds attest that Babylonians of all classes washed frequently and valued personal hygiene (Leick 2003:p.137-141). Many upper-class homes had dedicated bath rooms and built-in toilets; the less affluent could wash themselves in the canals that crisscrossed the Mesopotamian alluvial plain. Body cleaning was a matter of courtesy (visitors were provided with a basin to wash their feet when entering another person's house), sanitariness (people washed their hands before eating), and aesthetics (women bathed and anointed themselves with perfume before romantic interludes). The washing of the skin and other grooming procedures were also carried out in the context of certain medical treatments and religious rituals.

The five major episodes of body cleaning in the epic revolve around four activities: the bathing of the body, the grooming of the hair, the use of perfumed oils or other fragrances to scent the skin, and the wearing of fresh garments (Table 1.[3]). The one element that is common to all five examples is the donning of clean clothes. Where body cleaning is mentioned only in brief in the three passages involving Aruru, Enkidu and Gilgamesh, the key activity is the washing of the skin with water. Although there is no one characteristic that is common to all eight instances, they may nonetheless be grouped together into the

3 Episodes of body cleaning are arranged in the order in which they appear in the epic narrative.

	Bathing of Body	Washing of Hair	Scenting with Perfume	Dressing in Clean Clothes
Aruru Tablet I 101-102	X			
Enkidu OB Penn. 106-118	X	X	X	X
Ninsun Tablet III 35-46	X		X	X
Gilgamesh Tablet VI 1-9		X		X
Enkidu and Gilgamesh Tablet VI 167-168	X			
Gilgamesh Tablet XI 250-261	X	X		X
Gilgamesh Tablet XI 303-307	X			
Enkidu Tablet XII 31-36			X	X

Table 1. Elements of Body Cleaning in the Epic of Gilgamesh.

single category "body cleaning" using the principle of polythetic classification, which allows for items to be sorted according to family resemblance (Needham 1975). All that is required for an item to merit inclusion in a set is that it possesses a substantial number of the characteristics that define the set – it need not possess them all.

This multifaceted conception of body cleaning engages several sensory modalities, such as sight (replacing soiled garments with spotless ones); smell (perfuming the body); touch (by untangling matted hair); and temperature (by cooling the skin). Even kinaesthesia, the sensation of one's body in motion, is evoked in Ninsun's sevenfold entry into the bathhouse and the tossing of Gilgamesh's hair. Personal cleanliness is therefore conceived within the context of the epic as not only involving the whole of the body and its accoutrements, including hair and clothing, but also as implicating different types of sensations, including the visual, the olfactory, the tactile, the thermal, and the kinaesthetic.

Note that while some of the instances of body cleaning in the Gilgamesh epic refer to body hair, there is no definitive evidence for depilation as a fundamental aspect of personal hygiene in ancient Mesopotamia, as there is in ancient Egypt and the medieval and modern Middle East. Both razors and tweezers were in use by the late Early Dynastic period (ca. 2600-2350 BCE), although some scholars have theorized that the primary purpose of the tweezers was to remove lice rather than to pluck hair (Woolley 1955; Woolley 1934; Childe 1930:p.101). Shaving of the head, face, underarms and pubis is documented for magical, medical, and punitive purposes; partially or fully shaved heads and faces are also associated with certain classes of priests and temple functionaries, as well as slaves (Stol 2006-2008). In general, however, body hair is regarded favourably in the Mesopotamian written sources. For girls, the appearance of pubic hair represented the attainment of womanhood and sexual maturity, and love poetry extolled the hair surrounding a woman's vulva (Leick 1994:p.84, 96, 120). Copious hair was also considered 'a sign of primitive strength' (Foster 2005:p.99 n. 2; see also Foster 1977:p.80 n. 15).

The epic offers two grounds for body cleaning. One may wash after getting dirty or before undertaking an activity that requires absolute cleanliness. Gilgamesh and Enkidu rinse the blood and grime off their bodies after battle; both are also cleaned after wandering, unkempt, in the wilderness. On the other hand, Aruru washes her hands before sculpting Enkidu from clay, and Ninsun bathes before petitioning the god Shamash for the safety of her son.

It is notable that, in the context of the epic, these two distinct motivations for body cleaning break down along gender lines: it is the male characters who become dirty through physical exertion and heroic exploits, while the two female characters are never portrayed as soiled but only cleanse themselves in preparation for performing actions that will advance the interests of the male characters. This binary representation is in accordance with gender archetypes that, at least among the elite classes of Mesopotamian society, positioned men as active providers and political actors and women as passive homemakers and consorts. It is also significant that Aruru and Ninsun are goddesses, whereas Enkidu and Gilgamesh are mortal, if not fully human. The women's divinity may explain why the text does not depict them as sullied by earthly pollutants.

Two of the eight episodes offer no rationale for body cleaning; in these cases, the motivation may be pure enjoyment. Gilgamesh jumps into a pool of cool water after a long day of travel, and Enkidu dresses in clean clothing and anoints himself with oil on his foray into the Netherworld despite Gilgamesh's warnings. The sensory pleasures of personal hygiene are many, as outlined above – so much so that certain traditions of Christianity, for example, have at times rejected body cleaning as either a sinful indulgence or a carnal distraction from the more noble pursuit of perfecting the soul (Smith 2007). In the two examples in the epic, the desire for the gratifying sensations associated with physical cleanliness trumps both reason and self-interest, since, as a result of their actions, Gilgamesh loses the plant of youth he has worked so hard to obtain and Enkidu loses his very life.

The instances in the epic also differ with regard to who performs the action of body cleaning. In almost all cases, the character washes him or herself, but there are two instances where characters are washed by someone else. When Enkidu joins the shepherds' camp, he is first groomed by a barber before taking over the latter part of the cleansing process himself.[4] Similarly, when Gilgamesh leaves the home of Uta-napishti, he is led to a washtub and is assisted in bathing by the servant Ur-shanabi.

Physical Cleanliness as a Social Boundary

Body cleaning functions within the narrative to ready the individual for an assortment of social interactions. This is hardly surprising: the human body is always already culturally mediated, and a raft of social meanings attach immediately and automatically to its features (see also Kost, this volume). Cleanliness, variously defined, is an essential facet of self-presentation that has been used across cultures to distinguish insiders from outsiders, upper classes from lower classes. Classen, for example, has examined in depth how body odour – often interpreted as a sign of personal cleanliness, or lack thereof – has served as a marker of identity and difference in diverse cultural contexts (1993:p.79-105). Personal hygiene, moreover, is not just a state of being; it is an activity, a process that must be regularly repeated. The ritual nature of many cleaning behaviours provides another means of expressing membership in a social group (Yapici, this volume).

In the first tablet of the Gilgamesh epic, after Enkidu washes himself, he takes up arms against his former companions, protecting shepherds from wild animals instead

4 This interpretation rests on reading 'barber' rather than 'water' in OB Pennsylvania l. 106 (see note 1).

of defending wildlife from human encroachment. The act of body cleansing is part of Enkidu's acculturation, the final step to full acceptance into human society. The purpose of Ninsun's ablutions, on the other hand, are to ready her for ritual contact with a god. The first time Gilgamesh bathes, he finds himself the unwilling object of the goddess Ishtar's sexual advances. In every instance, the cleansing of the body provides entrée to a new social context: Enkidu gains admission to the company of other human beings, Ninsun to the company of the gods, and Gilgamesh to the intimate company of a woman.

When, at the close of the epic narrative, Gilgamesh bathes again in preparation for his return to Uruk after his quest to the end of the earth, his actions echo Enkidu's experience nine tablets earlier: once a solitary wanderer in the wilderness, the bather is now admitted – or readmitted – to the ranks of humanity and is fit to live amongst his brethren in the village or the city. In both cases, assistance is needed to initiate the cleaning routine. Enkidu, who has never learned how to bathe, and Gilgamesh, who has long neglected his personal care, are (re)introduced to the process like children.

The last major instance of body cleaning, which occurs in the epilogue to the text, is a sophisticated inversion of the examples of personal cleanliness that appear in the main epic narrative. When Enkidu cleans himself on his way into the Netherworld, the spirits of the dead identify him as a stranger. Rather than symbolizing conformity, his cleanliness represents an intolerable difference. Hygiene retains its social value: it is abhorrent to the phantoms precisely because it marks Enkidu as a member of human society, an order the dead stand beyond. Nonetheless, the episode concludes with Enkidu's admission to a new milieu, as he is seized by the Netherworld and the shades forcibly convert him into one of their own. Afterward, he describes his body as riddled with maggots and filled with dust (Tablet XII, ll. 96-99).

In every case cleaning is a transitional act, bridging states of being and transforming the socially unacceptable body into the socially appropriate body. Given that body cleaning as a practice is a threshold of social acceptability, it is therefore enlightening to observe who, in the epic text, is marked as unclean. The unclean fall into four categories:

A) The Wild Man

This, of course, is Enkidu, the newly-minted man, fashioned out of clay and dropped into the wilderness, the man who has never been part of society, who lives at a remove from both city and village. Enkidu is man in a state of nature, only recently created and not yet subject to the influence of civilization. As such, he is a stranger to both the practical procedures and the social values associated with personal cleanliness until they are taught to him. Prior to meeting other human beings, Enkidu lives dirty, naked, and carefree.

The relationship between cleanliness and nature – specifically, humanity in a state of nature – is not straightforward, however. After Enkidu has intercourse with Shamhat, his first human contact, his body is described as being 'defiled'. As a result, his animal friends become fearful of him, and, for his part, Enkidu loses some of his physical vigour and is no longer able to keep up with the herd:

> For six days and seven nights, Enkidu was erect and poured into Shamhat.
> After he was sated with her delights,
> He turned his face toward his beasts.

The gazelles saw Enkidu and drifted off.
The beasts of the wild shied away from his person.
Enkidu had defiled his body so pure;
His knees stood still, as his beasts moved away.

(Tablet I, ll. 194-200)

The difficult verb 'to defile' (*šuḫḫû*) that is used here refers particularly to debasement through sexual activity (Mayer 1988:p.157-158; Mayer 1997:p.177). The text therefore draws a contrast between dirtiness and moral impurity. Through his interactions with human beings, Enkidu may become physically clean but at the same time he corrupts his innocence.

B) Prostitutes

The next category of people portrayed as dirty in the epic are prostitutes. Enraged at Shamhat for setting him on a course that ultimately leads to his premature death, Enkidu curses her as follows:

[May you never] sit in the young women's [chamber]!
May the ground defile[5] your fine [garment]!
May [the drunkard] smear [your festive gown with dust]!

(Tablet VII, ll. 108-110)

What Enkidu describes here is the grim face of prostitution – furtive trysts in the dirt with drunks and rejection from polite society. In keeping with the epic's motifs of duality and reversal, however, after the sun god Shamash chastises him for blaming his misfortune on Shamhat, Enkidu compensates for his curse with a blessing that portrays, if somewhat sarcastically, a sex worker living the high life, lusted after by men of high rank who will richly compensate her with obsidian, lapis, and gold.

C) The Grieving

A third group of human beings whom the epic marks as unclean are people in mourning. As Enkidu lies dying, Gilgamesh swears to him:

And, as for me, after you are gone [I will let my hair grow matted.]
I will dress in a [lion] skin and [go roam the steppe.]

(Tablet VIII, ll. 90-91)

It was customary in Mesopotamia for mourners to express their grief by ceasing to care for their personal hygiene: donning sackcloth or other ragged clothing; refraining from bathing, grooming, or anointing themselves with perfumed oils; even smearing themselves with dirt (Pham 1999). The epic draws on these practices, which would have been familiar to a Mesopotamian audience.

While Gilgamesh's intentions are generally in keeping with traditional mourning customs, a deeper symbolism informs the specific terms of his pledge to Enkidu. Gilgamesh does not dress in rags to wander the streets of Uruk in despair. Rather,

---

5 This is the same verb, *šuḫḫû*, that refers to Enkidu's sexual defilement in Tablet I l. 199.

he clothes himself in animal skins and flees to the steppe. In essence, he becomes a wild man, the mirror image of Enkidu, and reverses the process of civilization that his friend experienced by exiling himself from human society.

Further, Harris has observed that Gilgamesh's uncleanliness shifts him into a liminal state – not alive or dead but 'betwixt and between' (2000:p.45). In this form, Gilgamesh has access to a kind of borderland, a mythological landscape inhabited by creatures that are neither human beings nor gods. This liminal status, like mourning itself, is a transitory condition that Gilgamesh cannot permanently occupy. Significantly, his vehicle for attaining this state is equally temporary: when he is ready to reclaim full membership in the human race, his skin may be washed, his hair combed, and his lion hide replaced with fresh clothing.

D) The Dead

The last class of individuals who are associated with uncleanliness in the epic have no reprieve from their piteous circumstances. After death, the corpse disintegrates and decays. Enkidu describes his body, after death, as 'filled with dust' (Tablet XII, l. 99) and devoured by maggots 'like an old garment' (ibid., l. 97). The spirit of the deceased finds little relief from this state of contamination in the afterlife. The epic refers to the Netherworld as the 'House of Dust' (VII 193):

> ...the house whose inhabitants are deprived of light,
> Where dust is their sustenance and their food is clay.
> They are dressed like birds in garments of feathers,
> And they cannot see light but dwell in darkness.
>
> (Tablet VII, lines 187-190)

All four groups have in common that they stand outside the social order. Each represents a different way in which one can be alienated from society:

- Enkidu, living in the wilderness, represents physical separation from other human beings
- Prostitutes represent economic or behavioural separation
- The grieving represent emotional or psychological separation
- The dead represent biological separation

The manner of exclusion also differs: the wild man has never participated in society; prostitutes have been cast out by society; and the grieving experience a self-imposed exile from society. The dead have undergone a transformation in their state of being and have fundamentally ceased to be human – for that reason, they can no longer be social actors.

Cleaning and Value

What then of the relationship between cleaning and value? What purpose does the motif of body cleaning serve in the narrative? The episodes of cleaning in the Gilgamesh epic involve several different kinds of value: social, spiritual, aesthetic, and even economic. Throughout the text, the clean body is portrayed as socially normative, a prerequisite to interaction with other human beings. Body cleaning is also linked to spiritual purity, as Ninsun undergoes ritual cleansing before communicating with the god Shamash. When Gilgamesh washes himself after his expedition to kill Humbaba, his clean body proves to be aesthetically pleasing to the goddess Ishtar, who is so moved by his beauty that she proposes marriage.[6] Finally, Enkidu's curse against Shamhat links uncleanliness with a reduction in the monetary value of sexual services. The dirty prostitute copulates with drunkards who presumably have little to offer by way of compensation, while the clean entertains dignitaries who reward her with gems and jewellery. No matter the value system, cleanliness is preferable to dirtiness.

While cleanliness is positioned as a virtue only implicitly in the Standard Babylonian Epic of Gilgamesh, it is recommended outright in a precursor to this version of the epic, where a character advises Gilgamesh:

> ... let your belly be full.
> You should be happy, day and night.
> Make merry every day.
> Dance and play day and night.
> Your clothes should be clean!
> Your head should be washed! You should bathe in water.
> Look down at the little one who holds your hand.
> Let a wife delight in your loins again and again!
>
> (VAT 4105, iii ll. 6-13)

One of the Sumerian Gilgamesh poems credits the hero himself with renewing ritual cleansing practices:

> Having brought down to the land the of Sumer, which by then had been lost since ancient times, the orders and the rituals, you put in order the rites of hand washing and mouth washing....
>
> (Death of Gilgamesh, *Me*-Turan, Segment F ll. 16-18)

The *me*s referred to here are a complex of divine properties or ordinances that constitute the norms of civilized life. Other Sumerian myths also count body cleaning among the *me*s. The story of Inana and Enki lists ritual handwashing among aspects of human civilization like wisdom, reverence, strife, the shepherd's hut, and the kindling of fire (l. 75).

Virtue though it may be, personal hygiene is also depicted in the epic as a potential liability. In the last two episodes of body cleaning, physical cleanliness and its accompanying pleasures come at the highest expense. While Gilgamesh is enjoying a bath in a cool pond, a snake comes along and steals the plant of rejuvenation given to him by Uta-napishti in consolation for failing in his quest for eternal life. As a result of cleaning himself on his way into the Netherworld, Enkidu becomes trapped in the afterlife. In both cases, death triumphs – snakes are associated with the underworld in Mesopotamian mythology – and life is lost due to the desire for purity (Leick 2003:p.141). Thematically, these outcomes support the epic narrative's thesis that immortality is not only an impossible but an inappropriate ambition for human beings, who should instead make the most of earthly pleasures like family, friendship, sex, and, indeed, personal cleanliness.

---

6 See Liebelt, this volume, for cleanliness as beauty in the modern Middle East.

The Babylonian Epic of Gilgamesh is one of the earliest literary explorations of the purpose and meaning of human life. The merit of bodily cleanliness and its indispensability to communal living are among the key learnings that Gilgamesh brings back from his journey of discovery – a piece of knowledge from before the flood. The hero embarks on a quest for immortality, only to discover that, for humankind, eternal life lies in enduring works and contentment in domestic comforts. The epic establishes the clean body as a prerequisite to social participation and, consequently, as necessary for human fulfilment. In a story intimately concerned with the inevitability of death, cleanliness emerges as a fleeting state, only achievable in life and therefore for us to enjoy while we may.

Bibliography

Childe, V.G. (1930) *The Bronze Age*. Cambridge, Cambridge University Press.

Classen, C. (1993) *Worlds of Sense: Exploring the Senses in History and Across Cultures*. London, Routledge.

Edzard, D. O. (1997) *Gudea and His Dynasty*. The Royal Inscriptions of Mesopotamia: Early Periods Volume 3/1. Toronto, University of Toronto Press.

Foster, B. R. (1977) Ea and Ṣaltu. In: de Jong Ellis, M. (ed.) *Essays on the Ancient Near East in Memory of Jacob Joel Finkelstein*. Memoirs of the Connecticut Academy of Arts & Sciences 19. Hamden, CT, Archon Books, pp. 79-84.

Foster, B. R. (2001) *The Epic of Gilgamesh*. New York, W. W. Norton & Company.

Foster, B. R. (2005) *Before the Muses: An Anthology of Akkadian Literature*, 3rd ed. Bethesda, MD, CDL Press.

George, A. R. (2003) *The Babylonian Gilgamesh Epic: Introduction, Critical Edition and Cuneiform Texts*. New York, Oxford University Press.

Harris, R. (2000) *Gender and Aging in Mesopotamia: The Gilgamesh Epic and Other Ancient Literature*. Norman, University of Oklahoma Press.

Hawthorn, A. (2015) "You Are Just Like Me": The Motif of the Double in the Epic of Gilgamesh and the Agushaya Poem. *KASKAL: Rivista di storia, ambienti e culture del Vicino Oriente Antico*, 12, 451-466.

Lackenbacher, S. (1971) Note sur l'ardat-lilî. *Revue d'Assyriologie et d'archéologie orientale*, 65 (2), 119-154.

Laessøe, J. (1955) *Studies on the Assyrian Ritual and Series* bît rimki. København, Ejnar Munksgaard.

Leick, G. (1994) *Sex and Eroticism in Mesopotamian Literature*. London, Routledge.

Leick, G. (2003) *The Babylonians: An Introduction*. London, Routledge.

Mayer, W. R. (1988) Ein neues Königsritual gegen feindliche Bedrohung. *Orientalia* Nova Series, 57 (2), 145-164.

Mayer, W. R. (1997) II. Babylonisch: Akkadische Lexikographie: CAD Š2 und Š3. *Orientalia* Nova Series, 66 (2), 165-180.

Needham, R. (1975) Polythetic Classification: Convergence and Consequences. *Man*, New Series, 10 (3), 349-369.

Pham, X. H. T. (1999) *Mourning in the Ancient Near East and the Hebrew Bible*. Journal for the Study of the Old Testament: Supplement Series 302. Sheffield, Sheffield Academic Press.

Smith, V. (2007) *Clean: A History of Personal Hygiene and Purity*. New York, Oxford University Press.

Stol, M. (2006-2008) Rasieren. In: Streck, M. P. (ed.) *Reallexikon der Assyriologie und Vorderasiatischen Archäologie*, 11. Berlin, de Gruyter, pp. 260-262.

Woolley, C. L. (1934) *Ur Excavations Volume II: The Royal Cemetery: A Report on the Predynastic and Sargonid Graves Excavated between 1926 and 1931*. London, British Museum/Philadelphia, Museum of the University of Pennsylvania.

Woolley, C. L. (1955) *Ur Excavations Volume IV: The Early Periods: A Report on the Sites and Objects Prior in Date to the Third Dynasty of Ur Discovered in the Course of the Excavations*. London, British Museum/Philadelphia, Museum of the University of Pennsylvania.

Section Five

Social Practices and Politics

Shaking out the Tablecloth – Uzbek Hospitality and the Construction of Boundaries of Belonging

Sebile Yapici

Cleaning up in hospitality contexts – An outline

I will show how cleaning around food preparation and hosting in Uzbekistan and the US-Uzbek diaspora is a vehicle for social bonding and the integration of new female family members into the household. It also builds on the archetype of the guest as outsider. Whilst hospitality has a special place in Uzbek culture and makes 'things to do with food' a vivid and intriguing area of ethnographic research, cleaning and its integrative social function for the family are the inward-looking part of a social practice that, at first glance, seems to put the visitor first. But at closer examination, the practice of serving food and cleaning up creates an inside and an outside which helps the family to maintain established (gender) roles and turn outsiders into insiders over time. I will lay out my argument by focusing on two aspects of cleaning in the Uzbek context. First, I will look at cleaning as a socially and communally experienced practice that is traditionally reserved for female kin and younger women of the neighborhood. Cleaning serves as a mediator for sustaining relations between neighbors and family members. I will describe the importance of food sharing in Uzbekistan to give more context and to illustrate what makes hospitality particularly relevant for the making and unmaking of social relations here. Then I will switch my focus to cleaning as the 'hidden side' of cooking and food sharing. I will relate these observations to a reading of cleaning practices as identity markers and on how the notion of cleanliness itself differs from society to society, from group to group, and even from household to household within one cultural context or locality.

Wherever food is prepared, somebody will have to clean up. This remains true, no matter whether it be in the form of casually making oneself a quick bite, cooking a representative meal for guests at home or commercially offering food at a restaurant. Whilst it is possible to eat in company or solitude, the same goes for cleaning. The washing of dishes and the tidying up of eating and cooking spaces may be a paid activity performed by staff or by a house cleaner. It may be done by those inhabiting the home, either as a chore they have to perform for themselves, or as a service to other members of the household. In that sense, cleaning

in: Bredenbröker, I., Hanzen, C. & Kotzur, F (eds) 2020: *Cleaning and Value. Interdisciplinary Investigations*, Sidestone Press (Leiden), pp. 183-194.

belongs to the realm of care practices. Daniel Miller (1998) makes the case that shopping for the family is care giving, because a lot of thinking and knowledge about every family member's preferences and needs goes into it. In the Uzbek case, care work can be divided into different tasks for every gender, shopping being the task often done by men, while cleaning is performed by women. This can change in the Uzbek North American diaspora, sometimes enforcing, sometimes evening out the traditional gender roles. Cleaning however may also, as I will show, function as an activity which makes it possible to work one's way into kinship status, to share experiences made whilst working hard as a form of bonding and to prove success in performing one's appropriate role in the Uzbek family.

Within the anthropology of food, the practice of cleaning is under-interrogated. There may be a reason for this. Often, cleaning does not happen where the eating happens, and food preparation may take place at a different time than tidying up. In the case of cleaning as part of Uzbek hospitality, cleaning is not directly visible because – as domestic labour carried out by women – it is neither paid labour nor can it be regarded as a gift to household and family. There is a rich body of literature on female unpaid labor for example in Susan Strasser's book on American housework or in Marjorie L. DeVault's book on social organization of caring as gendered work (Strasser, 1982; DeVault, 1991). DeVault draws on similar, ideal image of the American woman as a caring person, deconstructs this image and analyzes the caring and feeding of the family as unpaid female labour. These studies draw attention to the problem that a woman's work is never finished and that the conception of the housework as a female domain did not change significantly with women entering the job market. In my article, I will shed light on the role of women in households of an extended family network in Uzbekistan and examine if and how this role changes through migration.

Even though the dishes are often the same on many occasions, hosting in Uzbekistan differs depending on the number of guests and the degree of familiarity between guest and host (the more distant the guests, the more refined the food). Guests can be family members or neighbors, more distant kin, friends and random foreigners (like me). Hosting does not require an invitation from the host. Guests who come over unexpectedly deserve the same attention and quality of food as invited guests for a feast do. While shopping is often the job of men, the preparation of the space, cooking, serving and cleaning afterwards are considered female work. The only exception in which men cook food is *osh palov*, a rice dish with carrots and meat, that is considered the most important national dish. *Osh palov* can be prepared by men, depending on the occasion and/or the region. Cooking takes place in the kitchen, while the guests are already present in the guestroom, hereby separating female household members who prepare the food from those who eat. If there are new brides in the family, most of the work – setting-up, cooking, and cleaning – will be done by them, whereas the most senior woman will sit with the guests. If there is no bride in the family yet, the daughters are responsible for the largest part of the work. Often, unmarried girls from the neighboring households are called to help. Before or right after the guest arrives, a clean tablecloth is spread out. Clean here means that the cloth has been shaken out and freed from crumbs of previous meals. Fruits, bread and nuts are always provided as soon as the guest arrives. During the visit the women bring new pots of tea, clear used dishes away, bring new food and refill the plates with fruits and nuts. The cleaning of the dishes takes place during the visit or afterwards, always away from where the guests sit.

Figure 1. The ethnographer at work in the kitchen (Photo: Sebile Yapici).

One morning I was invited to a family home in rural Uzbekistan to cook *manti*, steamed dumplings. When I arrived, I was led into the guest room where pieces of cake, biscuits, fruits, sweets and tea where already arranged on the table. We sat down and had breakfast together. After we finished eating, we moved outside onto a *chorpoya*, a sitting platform in the backyard of the home to prepare the dumplings. The hostess, two sisters and her mother joined us to make the *manti*. Everybody started doing something: cutting the vegetables and the meat, kneading the dough, bringing ingredients. I wasn't given any tasks. After preparing everything, we had to wait for the dough to rise. I was led into a room with a television where they switched on a Russian channel, and later a music channel. We chatted a bit, but I worried that they would start preparing the dumplings without me. When I was finally allowed to leave the room, the preparation was indeed already ongoing. Since I had specifically asked to participate, I was allowed to help form the dumplings, and it was taken with a big surprise that I was able to form them acceptably (fig. 1).

We put the dumplings into a steamer and the two younger sisters went to the kitchen to clean and tidy up and prepare a salad. The mother, the oldest sister, and I sat on the platform and had tea until the food was ready. We ate indoors in the guestroom. After lunch, the mother and the oldest sister went to the television room again and I was told to rest and sleep for a bit, while the younger ones cleared everything away.

Many visits as a guest to an Uzbek family home are structured like this. Even though this invitation included preparing the food together because I wanted to learn about food preparation, my role as an outsider was clearly keeping me out of the cleaning and cooking zone. Even as a woman, I was allocated the role of a 'guest' or 'outsider' rather than being included into the group of women who were cooking and cleaning.

Figure 2. Tabletcloth with food, ready for guests (Photo: Sebile Yapici).

Researching Uzbek hospitality

Since my work is primarily about Uzbek migrants in the United States, I have conducted fieldwork among them for a total period of nine months between March 2017 and October 2018, and I have spent over three months in Uzbekistan to study everyday practices back home. After the dissolution of the Soviet Union and the launching of the Green Card lottery in 1995, more and more Uzbeks entered the US. Many came to earn money so they could help their families at home, and many stayed, pursuing an education and a life in the US. Now, in most of the bigger cities of the US, there are substantial Uzbek communities, the one in New York City being the biggest of them. There are no certain numbers but the American Community Survey from 2016 lists approximately 27.000 Uzbeks in New York City and about 57.000 Uzbeks in the US in total.

Food sharing, both on everyday occasions and on holidays, is an important part of Uzbek hospitality. While researching Uzbek food culture, I became aware that food preparation and cleaning work for guests helps to build a strong bond between hosts and helpers. During my last visit to Uzbekistan, I was invited to visit a family I had already met in 2016. I came with the family's relatives and they weren't expecting us, yet as soon as we got there, the tablecloth was slowly filled with nuts, chocolates, fruits, *non* (Uzbek bread, baked in a clay oven called a *tandir*) and green tea (fig. 2).

The bride and the unmarried daughter of the household spent the whole afternoon cooking and preparing the food and cleaning up after every course. After the initial serving of sweets, fruits and bread – standard for every guest entering an Uzbek household – the daughter served a big plate of delicious, hot and crispy potatoes fried in cotton oil, giving them a fruity taste with a refreshing hint of bitterness. As the daughter told me, potatoes were the best they could offer on such short notice. A soup with meat, potatoes and carrots would have taken too much time, while a steaming plate of *osh palov*, consisting of rice, carrots and meat – even though considered the most important Uzbek dish – would not have been enough on its own (see Zanca, 2017:p.73). We ate the savory potatoes before digging into the rice and then finished with a refreshing dessert of honey and watermelon. Uzbek standards of hospitality are not satisfied by simply serving tea and sweets, nor by offering a single dish. Whenever possible, Uzbek hosts will attempt to provide a full meal with a starter and main course as well as fruits for dessert. The preparation and cleaning up of such a feast are an important part of kinwork consolidated in the process of serving the guest. A guest would be served a proper dish, preferably an *osh palov*, often alongside a smaller dish, such as soups or little dumplings. If the guest is in a rush, something quicker will be cooked. In order to provide guests with such elaborate hospitality, female family members share the work that goes into preparing and cleaning. If an unexpected guest arrives and ingredients are missing, a female member of the neighboring household will bring whatever is missing. The more a guest is socially removed from, or unfamiliar to the hosts, the more elaborate and varied the meal. Yet, even at its most refined, the principle stays the same: bread and tea at the beginning, then a first course such as a soup and a second course such as *osh palov* or *manti*, later sweets, melon (if in season) or other fruits and tea.

Cleaning work as gendered kinwork

In the context of Uzbek hospitality, cooking and cleaning is gendered work reserved for women. Men sit and talk with other male guests. Men only cook *osh palov* on special occasions such as wedding feasts or as a particularity of some regions and households. The cleaning work, however, remains an entirely female domain. For women, there is a great deal of cleaning to be done in the aftermath of the meal served to the guest of honor. Fruit plates are changed, teapots are refilled with fresh tea, full plates are brought from the kitchen, empty ones are taken away, trash is cleared away, dirty plates and cups are replaced with clean ones. Meanwhile, in the kitchen – which in villages in summer is usually outdoors – the female host and the younger female family members perform all the cleaning work related to serving food long before the guests arrive, as well as during the meal. The *kelin* (new brides) – daughters-in-law of the hosting family – are especially involved and engaged in this work (Sancak & Finke, 2007:p.174; Zanca, 2017:p.73). After everyone leaves, they remain in the kitchen, wash the dishes and clear everything away. As patrilocal residence is common in Uzbekistan, *kelins* usually live in the groom's household after marriage and have to do a lot of work in order to be considered a 'good *kelin*'. Depending on the size of the family, *kelins* usually do most of the housework until they grow older, give birth and have daughters and daughters-in-law to help them (Zanca, 2011:p.160; Sancak & Finke, 2007:p.171). Women complain about their husbands, their mothers-in-law and about all the work they have to do, but it is considered impolite to burden someone with one's own problems by complaining too much, so they keep working and regard it as the usual state of things. In return, women are praised for being good cooks and good housekeepers, and girls learn these tasks from an early age. During the meal described above, we sat together with the mother of the household, Muhayo, and her youngest grandchild, a girl of two months old.[1] She was loved and hugged by all of us. Muhayo's sister-in-law took the baby into her arms, lifting the baby towards her face in rhythmic intervals she would kiss it every time her tiny

1 Names are all changed.

face came close. Before every kiss she told her enthusiastically how beautiful and what a good girl she was, how in no time she would take care of the guests, would cook the food, serve the guests, sweep the floor, wash the dishes. It was as if she wanted to bestow those qualities and manners of behaviour upon the girl.

In a process of slow social ascendency within the family, new brides and young women can work their way towards a slightly less laborious position. Here, the figure of the guest and the figure of the *kelin* are interesting opposites, since they are both initially external to the family but receive very different treatments, have different intentions towards the family and serve a different purpose in the making and reproduction of the internal family hierarchy. The bride is in the process of becoming a family member and has already relocated, whilst the guest is affiliated but not resident in the household. A host will always try to convince their guests to stay until they have cooked something, but if the visit is too short there will be at least bread and nuts or fruits from which one should take a mouthful in order to not leave the house unfed (see Nalivkin & Nalivkina, 2016:p.82). The term *o'zbekchilik*, which can be translated as proper 'Uzbekness', refers to a moral code in which the honoring of the guest holds a central position, both for internal family organization and hierarchies as well as in relation to non-kin. This can be seen in proverbs like 'the father is great – the guest is greater than the father' (*Ota ulug' – otadan mexmon ulug'*). The guest, not wanting to become an internal part of the family, here serves as a blueprint against which the *kelin*, who does want to become part of the family, can prove herself. She serves the external representation of the family through hosting and thereby slowly loses her status as 'foreign' to the family. This comes at a price. *Kelin*s have to adapt to household standards of the groom's family and work hard to become valid family members. Learning the new standards of cleanliness is crucial to becoming part of another family. Working together – that is, not sitting in the guest room and being served, not participating in the conversations between the older female family members and the guests, but rather serving the guests – creates and maintains a special bond between the women who are working. Rumors and news can be exchanged, family matters can be discussed and evaluated. By showing their contribution to the family, daughters-in-law become part of the family.

Embodied knowledge and (un)consciously cleaning

I experienced Uzbek hospitality both as a guest at the tablecloth (I will explain below) whilst eating and chatting with other guests; and as a helper in the kitchen, following the process of serving guests and setting the table(-cloth). As a female researcher it was probably easier to be allowed into the kitchen, but I was always considered a guest, meaning I was supposed to be excluded from these processes even though I expressed interest in them. Most of the times I could only watch or help with little tasks, that are more pleasant in nature, such as filling dumplings or stirring the food. Even when I was visiting a family regularly, I was never allowed to wash the dishes or sweep the floor. In rural Uzbek homes, the tablecloth itself is essential because it is often laid out on the floor and functions as a table. Seating a guest at a tablecloth is the most important part of the ritual of hospitality in Uzbekistan. The tablecloth remains a central organizing object for all the available foods. It can therefore be seen as a symbol of the cleaning work that goes into serving food. And even while the food is eaten and being served, those at the table contribute to the tablecloth's status as a clean object. It is important not to step on or over the tablecloth. Crumbs are removed during the meal, often almost unconsciously, by women belonging to the household, often during a conversation. One edge of the cloth is tightly held and hit with the side of the hand which is then immediately transformed into a bowl to catch the flying crumbs. I've seen this particular movement in different households and consider it an embodied as well as partially subconscious manifestation of a habitus around cleaning and eating which is particular to Uzbek hospitality. Everyday practices like cleaning become embodied by watching and imitating them from childhood on – particularly for girls and women, who are expected to take active roles as cleaners. As the story of Muhayo suggests, being good at cleaning, cooking and serving guests is a desirable skill for women in Uzbekistan. That is why girls are encouraged to imitate, watch and help with these practices. Depending on the social expectations that come as consequences of a person's assigned gender, social class and family role, the ways in which such practices are taught and learned may differ enormously. The fact that cleaning practices are embodied points to them as being something that is done partially unconsciously and partially in order to create a sense of home and orientation, as Ghassan Hage (2010) describes:

> "*The feeling of familiarity is generated by a space where the deployment of our bodily dispositions can be maximized, where we feel in possession of what Bourdieu would call a well-fitted habitus. ... This sense of implicit knowledge implies spatial and practical control that in turn implicates the sense of security ... .*", p.418.

It is a way of being in a world where we have to learn techniques to such a degree of mastery that they become second nature and can recede from consciousness. It is the way we speak or walk or swim or run or cycle – in the country and in the city, in this country and that, in this region and that, as men and women – which marks things like our accent and our posture. This is where individual habits are socialized. Habituated styles of cleaning are one of the central ways in which the social world puts its fingerprints on our bodies and through which culturally specific, aesthetic expectations of cleanliness and filth are established.

Culturally different notions of cleanliness – Insiders and outsiders

As I have just shown, cleaning activities that occur around hosting and preparing food play an important role in defining the interior and exterior of a family. Membership, particularly in the case of the *kelin*, can be gained and worked for through cleaning and cooking. Women are traditionally the ones put in the position of the 'foreign intruder' coming into a family. As learned through the social context in the family, women are prepared for fulfilling the role of a new bride as well as instructing children and new brides how to cook and clean in a way that will gain them acceptance and membership in the family. The guest is always someone outside the family towards whom hospitality can be shown, which helps to create friendly ties but also strengthens the family's sense of an interior and an exterior. The same mechanism may, in the Uzbek example as well as on a more general scale, also be true with regards to a sense of national identity and belonging. Hospitality and the food that is served are an integral part of the techniques and standards of hosting, important to the identity of being Uzbek (see Zanca, 2011:p.119; Turaeva, 2017:p.94). Identity of course is both a marker of distinction from others who do not belong and a marker of commonality with those who do. And more of these techniques and standards become natural – that is, unconscious – the more they make the subjects feel like essentially different people, socially constructed but totally natural. Nothing demarcates that boundary more clearly than disgust. Sara Ahmed (2014) dedicated a chapter in her book politics of emotions to disgust. There she puts food at the core of disgust:

> *"Food is significant not only because disgust is a matter of taste as well as touch – as senses that require proximity to that which is sensed – but also because food is 'taken into' the body. The fear of contamination that provokes the nausea of disgust reactions hence makes food the very 'stuff' of disgust. Of course, we must eat to survive. So the very project of survival requires we take something other into our bodies. Survival makes us vulnerable in that it requires we let what is 'not us' in; to survive we open ourselves up, and we keep the orifices of the body open."*, Ahmed, 2014:p.83.

If we think something is disgusting or dirty, we have learned these evaluations somewhere. Most likely, the places where these moral and aesthetic categories are learned and passed on are the family and the community in which one is socialized. Our cultural backgrounds and other factors such as gender or class determine what we regard as clean and unclean, how and how often we clean our body, the house and so on. Embodied and partially subconscious categories like those of clean and dirty are difficult to overcome or become conscious of precisely because they feel natural. Ben Campkin and Rosie Cox (2012) develop the argument, that 'being clean' is perceived as a static, natural thing that we can spot easily. They state that:

> *"In the 'modern' West it is generally assumed that such practices, and acceptable standards of cleanliness, have been developed in a linear fashion in response to scientifically informed 'germ theory', which unquestionably explains what is hygienic or unhygienic ... However, when cleaning practices are examined closely we find that the daily processes intended to keep bodies, houses or cities free of dirt are not organized in direct relation to unquestionable scientific evidence. Scientific definitions of 'dirty' and 'clean' are produced within particular historical and cultural contexts, rather than standing as objective truths."*, Campkin & Cox, 2012:pp.1-2.

Or, as Mary Douglas (2003) puts it, 'there is no such thing as absolute dirt: it exists in the eye of the beholder' (p.2). The same could be said for questions of edibility and inedibility in relation to food. Like notions of dirt and cleanliness, our perceptions of what is edible and inedible can change, but some taboos are almost impossible to break. In fact, eating and preparing food as well as ideas of cleanliness overlap most directly in our effort to not incorporate anything dirty into our bodies. Telling people that one eats comestibles which they avoid can provoke disgust, as if one could become contaminated by impure food that has been incorporated. The evaluation of something as dirty can have various implications: it can refer to metaphorical dirt that is passed on to us through people who are not considered pure, such as in some cultures through menstruating women, or through people from lower castes in the case of Hinduism. It can refer to food that is not ritualistically pure, such as non-halal slaughtered meat or meat from impure animals.

Agreed standards of hospitality and cleanliness are the exterior against which the sense of national identity takes form and shape. So inevitably the question arises: how do Uzbeks construct the boundaries of their community in the USA, my main fieldsite?

Do Standards of Uzbek Hospitality and Hygiene Change in the Diaspora?

Whilst researching Uzbek migrants in the US, my first problem was how to find and approach Uzbeks living there. I had some contacts from Uzbek friends in Germany and Uzbekistan and got in touch with Uzbek community organizations, like the Uzbek American Association of Chicago. But the most productive way of getting in contact was through social media. In a group called 'Uzbeks in the USA', I made a post, describing my research and within the first hour I received 70 answers, among them several invitations to private homes, with the offer of cooking for me and letting me taste Uzbek hospitality. With the help of this social media post, I visited around twenty families in eleven cities throughout the US. On my first trip after travelling to the West Coast, to Chicago and Washington DC, I arrived in New York City, where I later stayed for seven months in total. In the families, I was presented with at times very different situations regarding household organisation. Some had adapted the traditional gender roles of a working father with the mother at home, responsible for raising the children and managing the household. In many cases, I met a working woman who was still responsible for most of the household chores on top of her paid labor. Girls were overall more trained for helping with cooking and cleaning, but many were also protected by the mother from getting involved too extensively so they could focus on school. In some families, fathers helped substantially with household chores and felt equally responsible for the children. In almost all cases, the absence of in-laws, besides the lack of child-care that comes with the presences of family, was perceived as good for the nuclear family life. Women felt the relief of not needing to conform to a different family standard and men felt more at liberty to help with housework without being judged by their family for not being man enough.

Cleaning itself did differ from the circumstances in a village house, but not so much from an Uzbek urban household. The rituals around hospitality and the serving of food remained unchanged: it started the visit with tea and different sweet snacks like dried fruits, sweets, cake or other baked goods, continued with a first course or the main dish, or several dishes at once and ended with tea and something sweet. Being in a different environment and more importantly not having the extended family around changes the way in which the outside and the inside is constructed. In general, families in the US are more focused on their nuclear family. I observerd that some Uzbek-American families integrated other Non-Uzbek Americans as close family friends and kin. Many families have Russian-speaking friends from the former Soviet Union who share a similar history and background.

Even though I was fed an entire week of delicious Mexican food in one family, most of my hosts made a point of serving their special recipe of *osh palov*, to taste 'true Uzbek culture'. While I thoroughly enjoyed eating *osh palov*, I also loved the rare occasions in which the hosts wanted to cook their favorite dish, which was mostly Uzbek as well. I, as a non-Uzbek, was a foreigner who was treated as a guest everywhere, but since I was also not American, we shared the status of not being born where we met. What brought me even closer to Uzbeks that I visited was the fact that I knew their country of origin and had spent time there, spoke their language and knew their culture. This helped me in getting access to the kitchen, but it did not change the desire of my hosts to serve me the best food possible and to prevent me from helping with the cleaning.

Conclusion

To conclude, we could claim that in Uzbek social contexts, cleaning forms part of the hidden hermeneutic of kinship which holds the community together. Cleaning stands at the core of sociality because common understandings of cleanliness set boundaries for the community. To help others clean, we need to share ideas of what constitutes cleanliness. Cleaning is a rhythmic, not fully conscious, but habitual, routinised activity and congenial to talking, confessing and complaining while performing it. Other than the fact that helping each other with cleaning work is absolutely necessary for holding feasts, it therefore also provides a person with a place in the community through communication and exchange of knowledge, through becoming informed and having a space to express one's thoughts. Having a space to express herself is a rare thing for an Uzbek bride who has all the work to do and often very little privacy because she lives with her parents-in-law or in close proximity to the extended family. Brides find the space to do so during their time spent cleaning and cooking with other female members of the household. Performing cleaning labor creates a level of intimacy amongst those who participate in it. My hypothesis is that, because it is under-appreciated work, its value exists on the register of proximate affect. By that I mean that the less value a task is given from the outside, the more value may be invested in it at the inside emotional level. This again leads to the core of sociality. The boundaries set by cleaning rules can be as material as the tablecloth that separates the food from the floor and from parts of the bodies which are considered unclean, like feet for instance. They can also be symbolic boundaries such as separating children from the common table, as they are always messing with the order and transgressing set boundaries.

Although some important work has been done on the topic, cleaning remains under-theorized in general, especially as an everyday technique of community building with all its boundaries. My goal in this essay has been to demonstrate that cleaning practices and notions of cleanliness are significant parts of the Uzbek community's identity. There is no cultural activity without cleaning work. Researching culture, in my case food culture, requires attention to the labor that is done around the cultural event, in which the preparation and consumption of food takes place, in order to fully understand both, the event and the context.

Bibliography

Ahmed, S. (2014) *The Cultural Politics of Emotions*. Edinburgh, Edinburgh University Press.

Alymbaeva, A. A. (ed.) (2017) *Food and Identity in Central Asia*. CASCA Volume II. Halle (Saale), Max Planck Institute for Social Anthropology, Department 'Integration and Conflict'.

Bourdieu, P. (1977) *Outline of a Theory of Practice*. Cambridge Studies in Social Anthropology 16. Cambridge, Cambridge University Press.

Campkin, B. & Cox, R. (eds.) (2012) *Dirt. New Geographies of Cleanliness and Contamination*. London, I.B.Tauris.
DeVault, M. L. (1991) *Feeding the Family. The Social Organization of Caring as Gendered Work. Women in Culture and Society*. Chicago, The University of Chicago Press.
Douglas, M. (2003) *Purity and Danger. An Analysis of Concepts of Pollution and Taboo*. Mary Douglas Collected Works 2. London, Routledge.
Duschinsky, R., Schnall, S. & Weiss, D. H. (eds.) (2016) *Purity and Danger Now. New Perspectives*. London, Routledge.
Hage, G. (2010) Migration, Food, Memory, and Home-Building. In: Radstone, S. & Schwarz, B. (eds.) *Memory. Histories, Theories, Debates*. New York, Fordham University Press, pp. 416-427.
Levenstein, H. (2002) The American Response to Italian Food, 1880 – 1930. In: Counihan, C. M. (ed.) *Food in the USA. A Reader*. New York, Routledge, pp. 75-90.
Miller, D. (1998) *A Theory of Shopping*. Ithaca, New York, Cornell University Press.
Nalivkin, V. P. & Nalivkina, M. V. (2016) *Muslim Women of the Fergana Valley. A 19th-Century Ethnography from Central Asia*. Edited by Marianne Kamp. Translated by Mariana Markova & Marianne Kamp. Bloomington, Indiana, Indiana University Press.
Sahadeo, J. & Zanca, R. (eds.) (2007) *Everyday Life in Central Asia. Past and Present*. Bloomington, Indiana, Indiana University Press.
Sancak, M. & Finke, P. (2007) Konstitutsiya buzildi! Gender Relations in Kazakhstan and Uzbekistan. In: Sahadeo, J. & Zanca, R. (eds.) *Everyday Life in Central Asia. Past and Present*. Bloomington, Indiana, Indiana University Press, pp. 160-177.
Speltini, G. & Passini, S. (2016) Cleanliness issues. From individual practices to collective visions. In: Duschinsky, R., Schnall, S. & Weiss, D. H. (eds.) *Purity and Danger Now. New Perspectives*. London, Routledge, pp. 162-17.
Strasser, S. (1982) *Never done. A History of American Housework*. New York, Pantheon Books.
Turaeva, R. (2017) The Role of Food in Identification Process: Examples from Central Asia. In: Alymbaeva, A. A. (ed.) *Food and Identity in Central Asia. CASCA Volume II*. Halle (Saale), Max Planck Institute for Social Anthropology, Department 'Integration and Conflict', pp.93-102.Zanca, R. (2011) Life in a Muslim Uzbek Village: Cotton Farming After Communism. Case Studies in Cultural Anthropology. Belmont, California, Wadsworth, Cengage Learning.
Zanca, R. (2017) The Nexus of Cuisine and Social Life among Rural Uzbeks of the Ferghana Valley. In: Alymbaeva, A. A. (ed.) *Food and Identity in Central Asia. CASCA Volume II*. Halle (Saale), Max Planck Institute for Social Anthropology, Department 'Integration and Conflict', pp.71-91.

The Cleansing of a Political System: Obliterations, Burials and 'Reuse' of Palaces and Seats of Power in Central Italy (Seventh-Fifth centuries BCE)

Robinson Peter Krämer

This paper deals with pre-Roman Italy from the seventh to the early fifth century BCE. During this period, tyrannies, autocracies and monarchies were overthrown in favor of republican systems of governance in the city-states. I argue that as the consequence of such an overthrow of governance, the public institutions of rulers' palaces and seats of power were cleansed by removing their previously central function for public life, framing them as symbols of the now disgraced systems and reusing them in very different ways. Members of early republican city-states in Central Italy used at least three main strategies for the cleansing of their political systems: (1) ritual obliteration; (2) destruction; and (3) the 'reuse' of palaces for different purposes.

Introduction

Many societies in human history have undergone an abrupt change from one political system to another, whether through revolution, social upheaval or military conquest. This political shift can, for instance, lead from a totalitarian regime to a democratic society or the other way around. It always, however, involves the following question: what happens to monuments, icons and symbols of the former regime?

Such symbols have often been the target of hatred and been destroyed, in the sense of a *damnatio memoriae*. One might think of the destruction of Saddam Hussein's statue in Firdos Square, Baghdad on April 9, 2003 (Farrell, 2008) or of the demolition of monuments of Lenin after the fall of the Soviet Union (a process known as 'Leninopad' – 'Leninfall'; Ion et al, 2014/2015). On the other hand, symbols and buildings of former regimes may also survive political change, as in the cases of Confederate flags in the United States or army barracks in Germany with name references to National Socialism. Sometimes symbols and buildings are even intentionally conserved and 'musealized', as in the case of the Berlin Wall after the reunification of Germany in 1990. These examples show the very different strategies of societies to deal with the unpopular past and its testimonies.

in: Bredenbröker, I., Hanzen, C. & Kotzur, F (eds) 2020: *Cleaning and Value. Interdisciplinary Investigations*, Sidestone Press (Leiden), pp. 195-212.

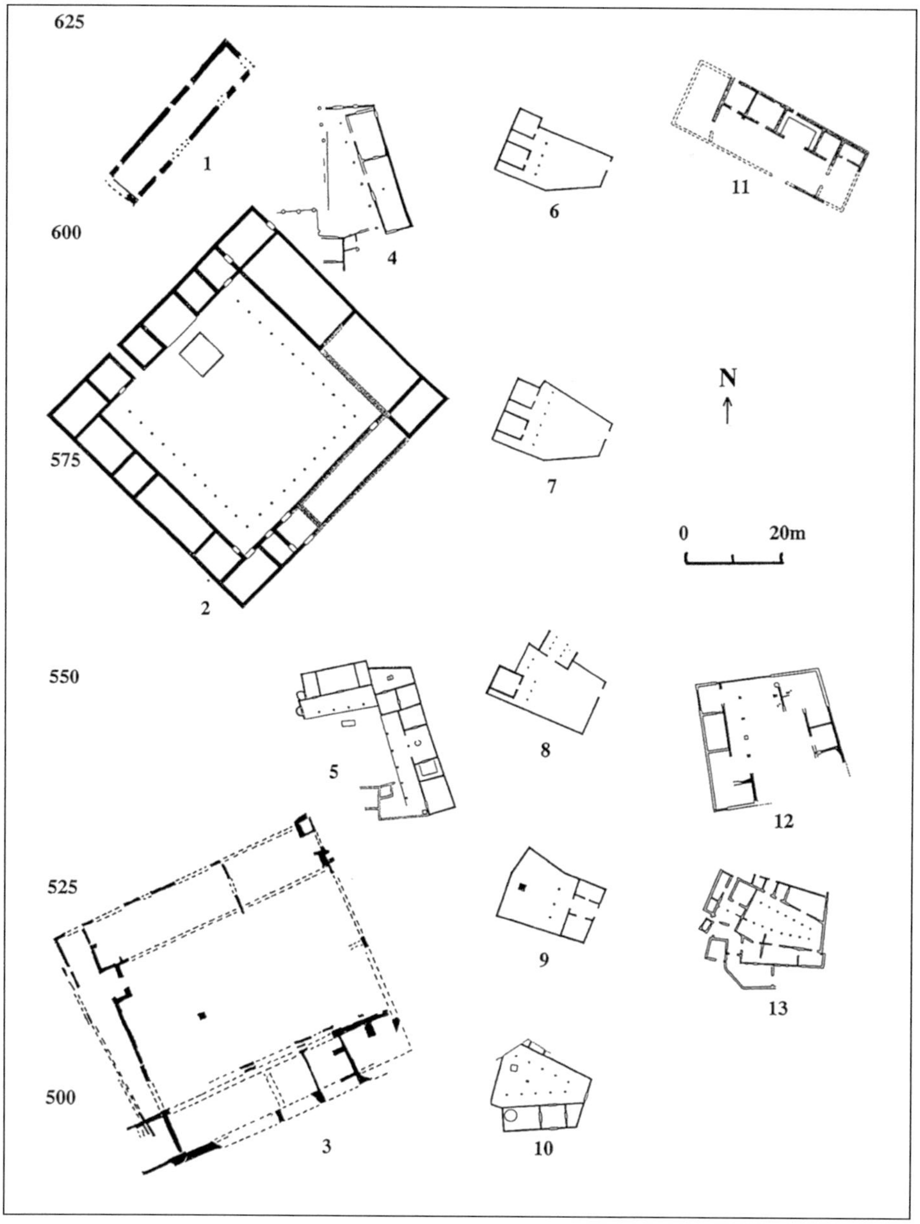

Figure 1. Archaic residences in Central Italy ordered by size and chronology. 1-2: Murlo, Poggio Civitate. Orientalizing and Archaic Phase; 3: Caere, Montetosto; 4-5: Acquarossa (phases 1-2); 6-10: Rome, *regia* 1-5; 11: Rome, *domus Regia*, phase 3; 12: Satricum, building A; 13: Athen, Agora. 'Building F' (Prayon 2010: 18 fig. 8).

This case study deals with Rome and the Etruscan city-states from the late seventh to the early fifth century BCE. While a city-state (Etruscan *spura*) was independent from others (as was the Greek *polis*), many were governed by autocratic rulers, tyrants or monarchs.[1] The details of the political organization of Rome and the different Etruscan city-states are unclear and subject to ongoing debate. The fact that the cities in Etruria and *Latium vetus* were ruled by autocrats, however, is widely accepted among scholars.[2]

A ruler's political power was represented by monumental architectural structures with luxurious equipment, rich palaces and seats of power (fig. 1; ancient Greek: *anaktora*; Latin: *regiae*). Palace complexes served as socio-political nexuses, even as sanctuaries, and were often located in the centre of cities and settlements or village clusters from the late seventh century to the beginning of the fifth century BCE (for examples and characteristics of *anaktora*/*regiae* in Central Italy, see Torelli, 2000; Marcattili, 2005; Prayon, 2010).

At the beginning of the fifth century BCE, most of the political systems built around dynasties and rulers seemed to have collapsed in a chain reaction of social upheavals. Monarchs and autocrats were killed or cast out of the city-states in a sudden disruptive wave of revolution, and the political systems were replaced by republican forms of government (Torelli, 1990; Cornell, 1995:pp.215-241; Maggiani, 2000; Aigner-Foresti, 2009:pp.14f.; Becker, 2013:pp.351-354; Tagliamonte, 2017:pp.128-130).

This article deals with the question of what happened to palaces and seats of power after the political changes. Were they destroyed, left intact or reused? The four case studies examined indicate that societies found different solutions and approaches to 'cleanse' new political systems of the monumental relics left by former regimes.

The *regia* and the Temple of Iuppiter Optimus Maximus in Rome

Probably the most famous, but also the most controversial, example of a palace or residence structure in Archaic Central Italy is the *regia* located in the Forum Romanum in Rome.[3] The *regia* was situated on the east side of the Forum Romanum, which had been the commercial, public, social and political centre of Rome most likely since the Orientalising Period, *i.e.* the early seventh century BCE. This seat of power has been typically interpreted as a part of a larger functional complex including sanctuaries, the official palace of the king and the houses of priests and the rulers.

The archaeological reconstruction, architectural context and many other aspects of the Roman *regia* are still a matter of intense discussion. For the purposes of this article, however, it is sufficient to observe that the *regia* in the Forum Romanum was a closed courtyard building with two or three rooms that was frequently rebuilt between the late seventh century and the end of the sixth century BCE (phases 1-4; fig. 1, no. 6-9). As a result, the *regia* frequently changed its structure and alignment without moving, however, from its central position on the *Sacra Via*, the main street for religious and triumphal processions.

---

1 For the organisation of Etruscan cities, see: Bizzarri, 2013; Menichetti, 2017; Sassatelli, 2017.

2 For Rome, see: Cornell, 1995:pp.119-150; Smith, 1996:pp.185-223; Aigner-Foresti, 2003:pp.125-138; Cifani, 2017b; for Etruria, see: Maggiani, 2005; Aigner-Foresti, 2009; Becker, 2013:pp.351f.; Tagliamonte, 2017; and for Central Italy in general: Glinister, 2006 with further references.

3 See, for example, Coarelli, 1983:pp.65-79; Cristofani, 1990:pp.47-76, esp. pp.59-61; Scott, 1999; Marcattili, 2005:p.310 Nr. 3; Murgan, 2011; Carnabuci, 2012; Potts, 2015:pp.140f.; Hopkins, 2016:pp.39-53 (cautious); van't Lindenhout, 2017:pp.88-92 (critical) with references to Brown and many other publications on this building complex.

Some pottery inscriptions suggest the reign of a king during the Archaic period with a residence in the *regia* or in the Forum Romanum: one Bucchero cup from the *regia* dating from the sixth century BCE (probably phase 4) shows the inscription *rex* (king), while the Lapis Niger in the northwestern part of the Forum Romanum contains a long inscription with the words *recei* (derived from *rex*) and *kalatorem* (from *kalator* meaning 'herald'; Hopkins, 2016:pp.48-53 with notes 45-46 and references).

According to written sources, the last king of Rome, Tarquinius Superbus, was overthrown in 509 BCE, marking the beginning of the Roman Republic. The archaeological record suggests that the *regia* burned down at the end of the sixth century/the beginning of the fifth century BCE, but was immediately rebuilt (phase 5; fig. 1, no. 10). Interestingly, only a few changes seem to have been made following this reconstruction. At the same time, the meaning and function of the *regia* was changed: now it served as a sanctuary and seat of high priests – the *rex sacrorum* and the Pontifex Maximus. It seems that the religious powers of the former king were transferred to a high priest and symbolic king, the *rex sacrorum*. According to Dionysius of Halicarnassus (4, 74), laws made sure that the lord of these rites would not use his power for political ambitions. The king of sacred rites was exempt from all military and civil duties, exercising the superintendence of the sacrifices *and nothing else*.

After the early Republican period reconstruction at the end of the sixth and the beginning of the fifth century BCE, the *regia* seems to have been rebuilt in the late third or early second century BCE with slight modifications. There are no archaeological traces of building activities after the fire of 148 BCE. The last complete rebuilding of the *regia* was made under Cn. Domitius Calvinus in 36 BCE after his victory in Hispania. Calvinus' *regia* essentially reproduced the architectural shape of *regia 5* from the end of the sixth century BCE and the beginning of the Roman Republic, while the Archaic *regia* (phases 1-4) frequently changed its layout. It seems clear that at the beginning of the Roman Republic a prestigious residence of a king was transformed into a public sanctuary with one definitive layout that kept its essential architectural form for centuries. In the case of the Roman *regia*, then, we can state that Roman society changed its symbolic significance from a seat of power to a Republican public temple – a complete shift of meaning in the public memory of Roman society.

Understandably, Elizabeth van't Lindenhout (2017) is deeply suspicious of the continued existence of a 'tyrannical' palace and symbol of oppression for centuries and has recently expressed her doubts about this interpretation and the understanding of the *regia* as part of a royal palace complex: 'Finally, it is remarkable that according to Brown's reconstructions and interpretations a royal residence was turned into an apparently important Republican monument: Regia 5. This building has been preserved and maintained during the Republic and Empire. Shouldn't we expect the Regia, the assumed royal residence, to be destroyed under the new Republican regime? Apparently, there were no hard feelings toward this building, on the contrary, it was maintained over eight centuries' (p.91). In the following, an attempt shall be made to show how this symbol of hated royal rule was nevertheless preserved and why this is not in contradiction to Brown's interpretation but rather makes sense as part of a broader social phenomenon in Central Italy.

The Temple of Iuppiter Optimus Maximus on Capitoline Hill offers an interesting parallel in respect to the change in the public memory of a building in Archaic/early

Republican Rome. Determining the exact shape, size and reconstruction of this building is very controversial as mainly parts of the temple podium and architectural elements in clay have survived.[4] Nevertheless, it was built in the sixth century BCE and was by far the largest temple of its time in Etruria and *Latium vetus*. Written sources state that Tarquinius Priscus began construction of the building, while Tarquinius Superbus completed it shortly before 509 BCE. Roman historians (Liv. 1, 55, 1-2; 2, 8, 5-6) underline the fact that the cult site was completed by the king, but remained undedicated when he was deposed (*'nondum dedicata erat in Capitolio Iovis aedes'*; Liv. 2, 8, 5). One of the first acts of the early Republic would have been the dedication of this highly important cult place. Suspiciously often, Roman sources highlight the tale of the powerful, tyrannical king of Rome, who was able to build the largest temple in Central Italy, but (luckily for the young Roman Republic) could not dedicate this extremely important and prestigious building. Instead, one of the very first actions of the newly founded Roman Republic would have been the dedication of the Temple of Iuppiter Optimus Maximus (Liv. 2, 8, 6-9). I suggest, that the tale of this dedication is overly loaded with symbolism, but not very plausible, and could point to the possibility of two dedications of the Temple: the first by the king, and a second time shortly afterwards at the beginning of the Roman Republic. Since this building was sacred and simply too expensive to destroy for being a symbol of monarchy, the Romans instead chose a new (Republican) dedication. The second dedication altered the public memory of the temple and transformed it from a symbol of an autocratic regime to a Republican building.

The *regia* of Gabii

A strikingly different way of dealing with the cultural memory of a palace can be observed in Gabii. This Latin settlement was situated to the east of Rome, along the Via Praenestina and on the now drained Lago di Castiglione. The settlement existed at least since the beginning of the early Iron Age and its heyday was during the Orientalising Period. Gabii exhibits interesting parallels with other Latin Iron Age settlements and especially with early Rome. The *arx* was located in the northeast of the settlement, where a *regia* was built that shows strong similarities to the one (especially to *regia* 4) at the Forum Romanum in Rome (Fabbri, 2015; Fabbri & Musco, 2016; Fabbri, 2017).

The earliest features of the *regia* are rectangular hut structures (approximately 6 x 8 m.) from the ninth and eighth centuries BCE. Post-holes and ditches are cut into the tufa rock and still preserved. In the second half of the seventh century BCE, the Orientalising period and monarchical phase at Rome, a stone building complex occupied the *arx* (fig. 2a). Its walls are made of stones of irregular shape and dimensions, but the exact form of the building is not clear. Nevertheless, it seems to have been a large, public building with a large courtyard oriented in the direction of north-west to south-east. Marco Fabbri interprets the building as being highly prestigious, not only because of its architectural ground plan and high-quality design, but due to the storage jar sherds found on site with fine 'white-on-red' decorations and early Latin inscriptions. This is a very unusual feature for *Latium vetus* in the seventh century BCE (Fabbri, 2015:pp.187-189; Fabbri, 2017:pp.226-229).

4 For the intensive discussion and the state of the debate, see: Mura Sommella, 2009; Hopkins, 2012; Potts, 2015:pp.123f.; Hopkins, 2016:pp.97-125; Cifani, 2017a.

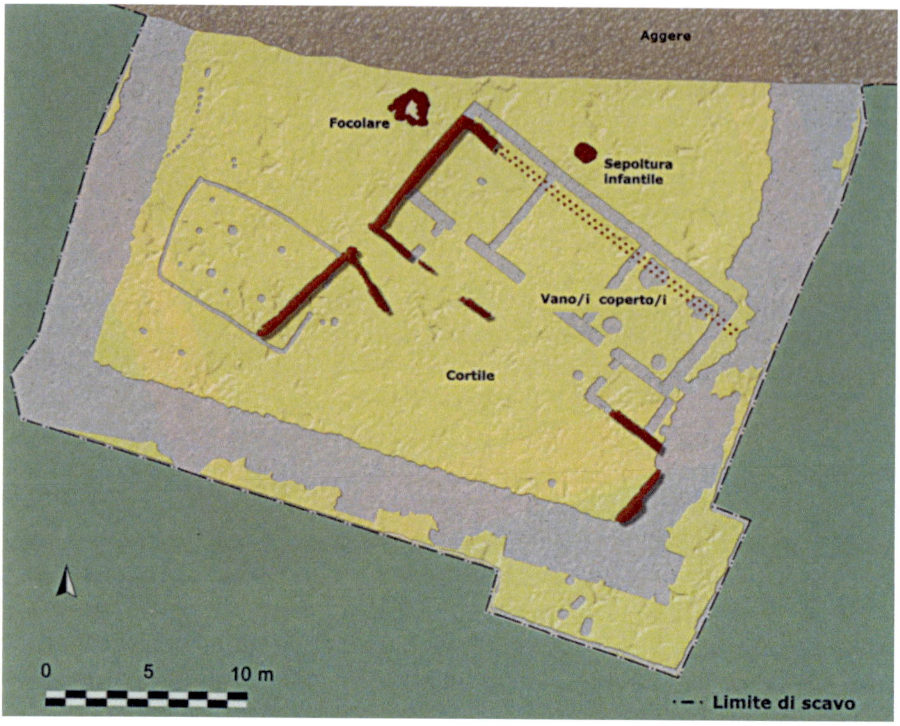

Figure 2a. Gabii, *arx*. Residence of the Orientalizing period (Fabbri, 2017: 227 fig. 3).

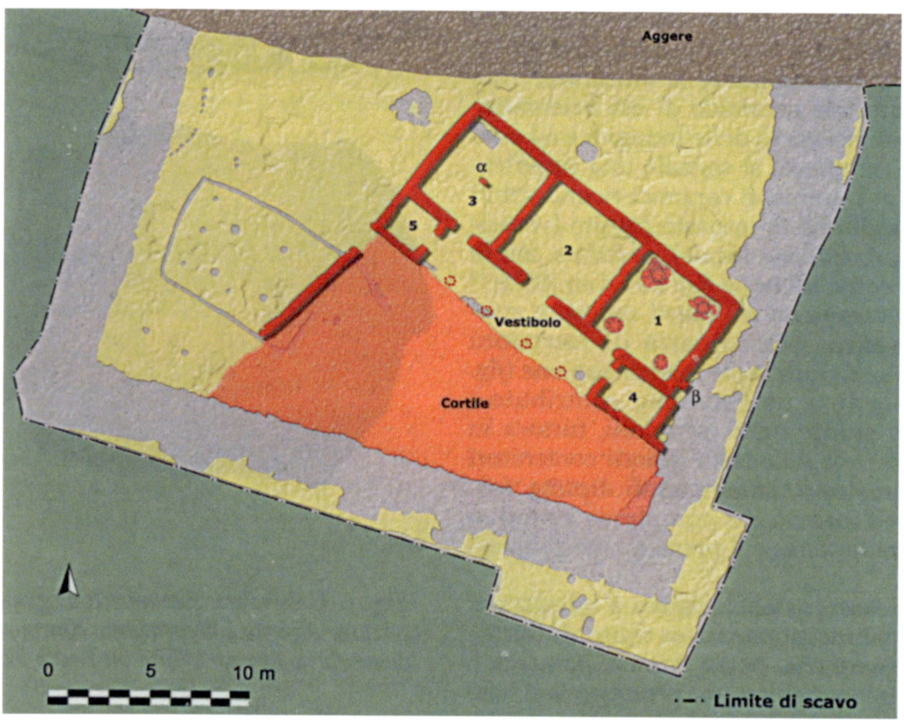

Figure 2b. Gabii, *arx*. Archaic *regia* (Fabbri, 2017: 227 fig. 4).

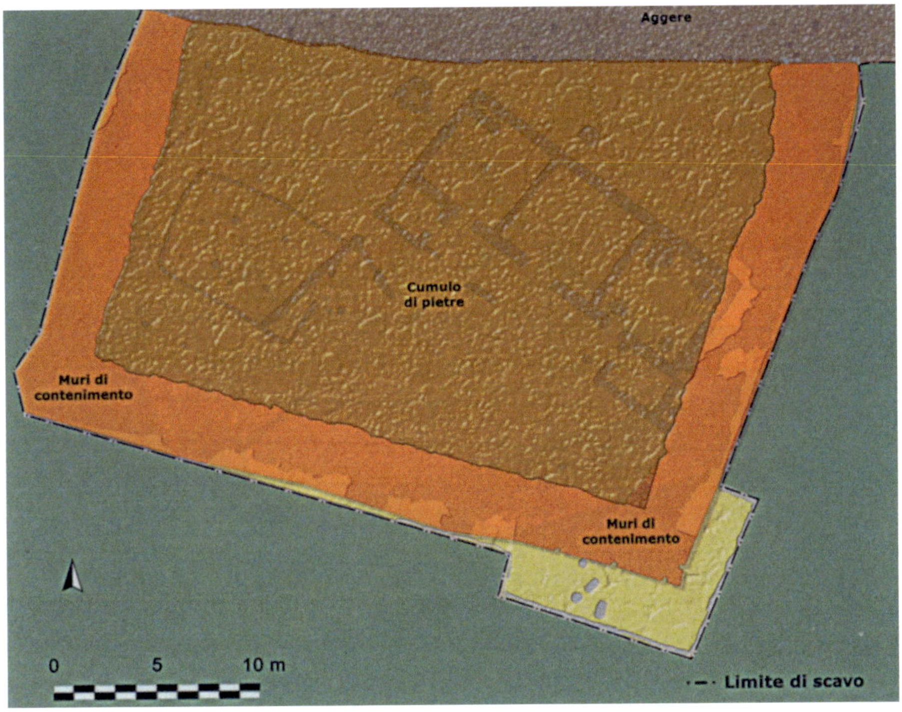

Figure 2c. Gabii, *arx*. Stone mound with wall structures (Fabbri, 2017: 228 fig. 5).

At the beginning of the sixth century BCE, a new complex was built that re-utilised some walls of the earlier building (fig. 2b). This courtyard building is very imposing, even monumental (western side: 20 m.; eastern side: 26 m.) and consists of a three-room complex with a vestibule and a courtyard with a similar orientation to the previous Orientalising building. The *regia* of Gabii, with its tripartite complex, strongly resembles the shape of *regia* 4 in Rome and of other *anaktora/regiae* (Fabbri, 2015:pp.189-196; Fabbri, 2017:pp.229-232). While the parallels are striking, it is even more interesting to observe how different its fate was from that of the Roman *regia*.

From the end of the 6th century to the beginning of the 5th century BCE, at the time of the political upheaval in Rome and parts of Etruria, the building was ritually deconstructed and obliterated (fig. 2c). The floors and upper walls were intentionally removed with the remaining structure carefully demolished and buried in a gigantic stone mound (at least 4 metres high), contained by wall structures on three sides. The site's excavator, Marco Fabbri, stresses how carefully this systematic process of obliteration was planned (Fabbri, 2015:pp.191f. 196; Fabbri, 2017:pp.232-236).

Murlo, Poggio Civitate

The ritual burial and demolition of a public building can also be traced at Murlo, our next example. Poggio Civitate is situated near the modern village of Murlo in Northern Tuscany, in a 'geo-political corridor' between the Etruscan city-states of Volterra, Vetulonia and Chiusi. For this reason and, due to its strategic position in the Ombrone Valley, archaeologists tend to interpret Poggio Civitate and its cluster of settlements in a political corridor as having a gateway function.

During the seventh and sixth centuries BCE, the hilltop of Poggio Civitate was occupied by a residential structure for members of a powerful social elite.[5] The beginning of this seat of power can be dated to around 680/670 BCE. Recent excavation campaigns have shown that there were at least five buildings: a large building (Orientalising Complex Building 1) interpreted as the actual residence; a fifty-two metre long pavilion structure (OC2) that served as a workshop with many manufacturing activities such as bronze casting, bone carving, pottery and terracotta production, as well as textile manufacture; a tripartite building (OC3) with a central room and two side chambers along with banqueting pottery. Excavators have usually considered this building as a banquet hall, but recently it has been instead interpreted as a temple. The purpose of the recently discovered buildings OC4 and OC5 remains unclear (Tuck et al, 2016).

Around 600 BCE, a large single fire destroyed the entire complex abruptly – probably by accident. In any case, the inhabitants constructed a massive four-winged courtyard building immediately afterwards, each wing measuring sixty meters long (fig. 1, no. 2). Like the Orientalising building, this palace was decorated with terracotta sculpture and plates. The frieze plaques depict processions of chariots, seated noblemen and women adorned with symbols of power, banquets and horse races; that is, the ideals and ideology of the social elite ruling this place.

This building was in use until ca. 530 BCE when it was destroyed and abandoned. At that time, the building was dismantled, the statuary and terracottas removed, smashed and ritually buried in pits and wells (Edlund-Berry, 1994; de Grummond, 1997; Tuck et al, 2010; Tuck et al, 2016:pp.87-108, especially pp.105-108). After this ceremonial destruction and 'un-founding' the palace was never rebuilt or used again.

Caere, Vigna Parrocchiale

The last example shows yet another different approach in dealing with a palace or seat of power. Caere (Etruscan *Kaisraie/Kaisrie*, Latin *Caere*, Ancient Greek *'Αγυλλα*) was one of the richest and most important Etruscan city-states due to its fertile soil (Liv. 28, 45, 14-15; Mart. 13, 124) and access to metal resources in the nearby Tolfa mountains and to Mediterranean long-distance trade. The city of Caere was situated on an approximately 160 hectare plateau between the rivers Manganello and Mola.[6] At the heart of the plateau settlement lies the archaeological area of Vigna Parrocchiale with two main phases (Cristofani, 1992; Cristofani, 1993; Moretti Sgubini, 2001:pp.121-141; Cristofani, 2003; Bellelli, 2008).

In a first building phase (6th century BCE; fig. 3a-b), a monumental palace or residence (*'residenza arcaica'*) was erected. It may have belonged to Thefarie Velianas, the king or autocrat who dedicated a large part of the port of trade and harbour sanctuary of Pyrgi (Cristofani, 2003:pp.69-246; Krämer, 2016:pp.81f. with fig. 5-6 and references). The inscription *mi calaturus* ('I belong to the herald')[7] on a pot sherd from the residential complex suggests the regular presence of a herald in the palace. A cistern and the remains of bronze processing near the monumental complex belong to the same period of use.

5 For the still ongoing excavations of Poggio Civitate see Tuck & Wallace, 2013; Tuck, 2014; Tuck, 2015; Tuck et al, 2015; Tuck et al, 2016 with further references.

6 For Caere see Cristofani, 1987; Maffei & Nastasi, 1990; Colonna, 2010; de Grummond & Pieraccini, 2016.

7 Pandolfini Angeletti, 1989:pp.70; 72 no. 1; 73 fig. 1; Cristofani, 1992:p.150 no. E 17.3; Moretti Sgubini, 2001:pp.121. 129 no. II.A.2.15; Aigner Foresti, 2009:p.11.

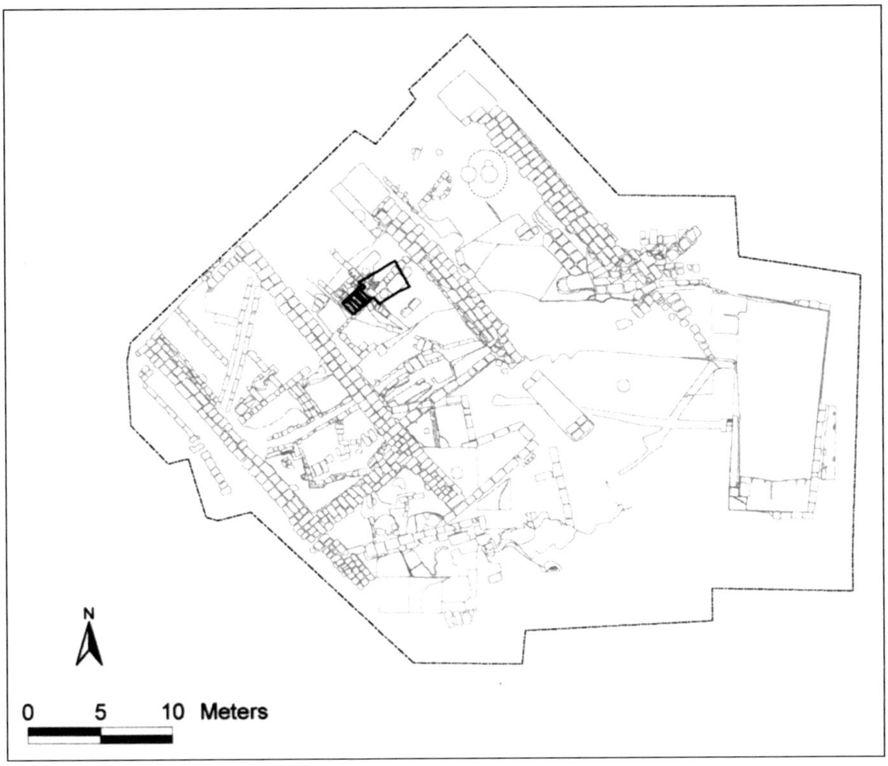

Figure 3a. Caere, Vigna Parrocchiale. Archaic residence, phase 1 (courtesy of the CNR – ISMA Archive).

The residence was in use throughout the sixth century BCE until the end of the sixth and the beginning of the fifth century, when the entire area was destroyed and the rubble ('*scarico arcaico*') dumped into the cistern (Cristofani, 1992; Cristofani, 1993; Bellelli, 2008:pp.323-325). Shortly afterwards, around 490-480 BCE, the former palace area was replaced, or even 'sacralised', by a monumental Tuscan temple (fig. 3c; ca. 24,5 x 20 m.) that was built directly on top of the former residence. At the same time, a large building with an elliptic shape ('*edificio elittico*'; ca. 22 x 32 m.) was constructed nearby, which has been interpreted as an *ekklesiasterion*, a public building for meetings in a similar sense to a town hall.

Cleansing Practices of Political Systems? The Social and Political Context

How can we interpret these case studies of seats of power and what do ancient sources state about transformations, obliterations and ritual burials of places? In the case of the Roman *regia*, the palace building was reused, even rebuilt, for different purposes – as a sanctuary and home of a high priest (*rex sacrorum* and *pontifex maximus*). The conservation of the building's basic architectural design for many centuries, until the Roman imperial period, may be characterised as a conservation, or even the 'musealization' of the former seat of power. The cases of Murlo, Poggio Civitate and the *regia* of Gabii – that is, the obliteration and 'un-founding' practices directed to both palaces – point to the ritual elimination of residences from

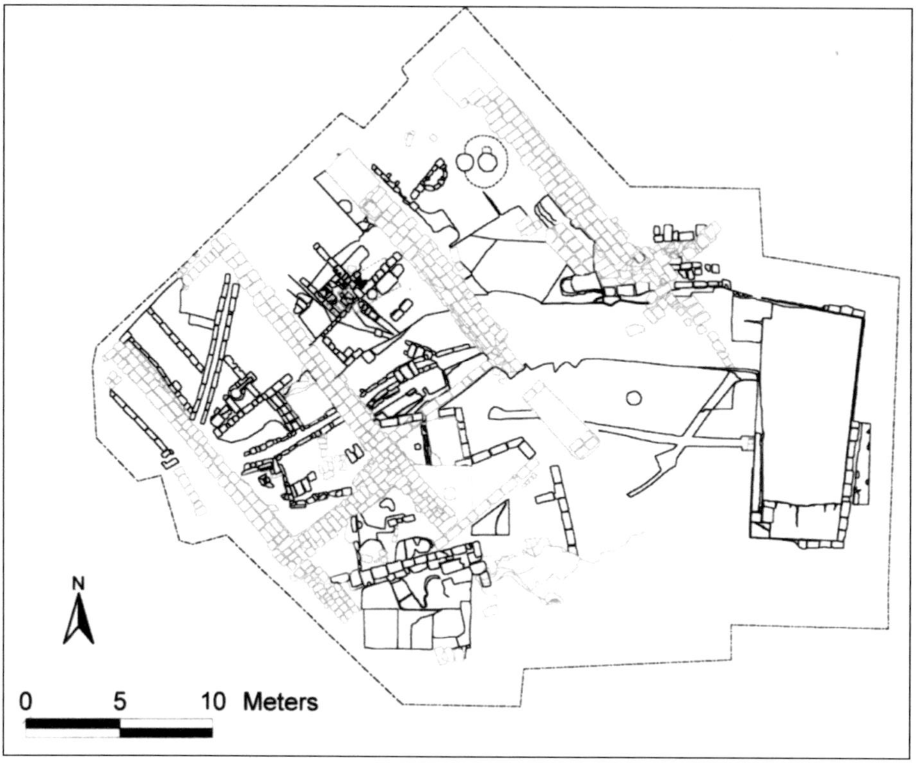

Figure 3b. Caere, Vigna Parrocchiale. Archaic residence, phase 2 (courtesy of the CNR – ISMA Archive).

the physical landscape and public memory. The *regia* of Gabii was dismantled, filled with large rocks and covered by a large tumulus, while the *anaktoron* of Poggio Civitate was carefully disassembled, with the architectural elements being ritually buried in pits and wells. The palace of Vigna Parrocchiale in Caere was simply destroyed and replaced by a monumental temple and an *ekklesiasterion* – a new meeting place for society at the dawn of the Republican era. Traces of the old regime were here banished from public memory.

Further archaeological evidence: Vetulonia and Poseidonia

Other archaeological examples in Italy might help us to understand ancient societies' approaches to cleansing their difficult pasts and political systems. Such a case study is the helmet deposit on the *arx* of the Northern Etruscan city-state of Vetulonia (most recently Maggiani, 2012; Pucci, 2016; Amann, 2017:pp.988f.; Egg, 2017:pp.172f. with further references). The deposit consists of almost 150 helmets[8] of the type Negau, variant Vetulonia, that were ritually deposited on the *arx* of Vetulonia in the first half of the fifth century BCE (probably ca. 470-450 BCE). The helmets were intentionally crushed, smashed and damaged (probably with axes or similar weapons) and partly inserted into one another before the ritual destruction. The damage can therefore not have been the result of acts of war. About sixty of the helmets bear the inscription *haspnaś* – belonging to (the family,

8 Not ca. 125 helmets as often written by scholars, see Maggiani, 2012:p.63.

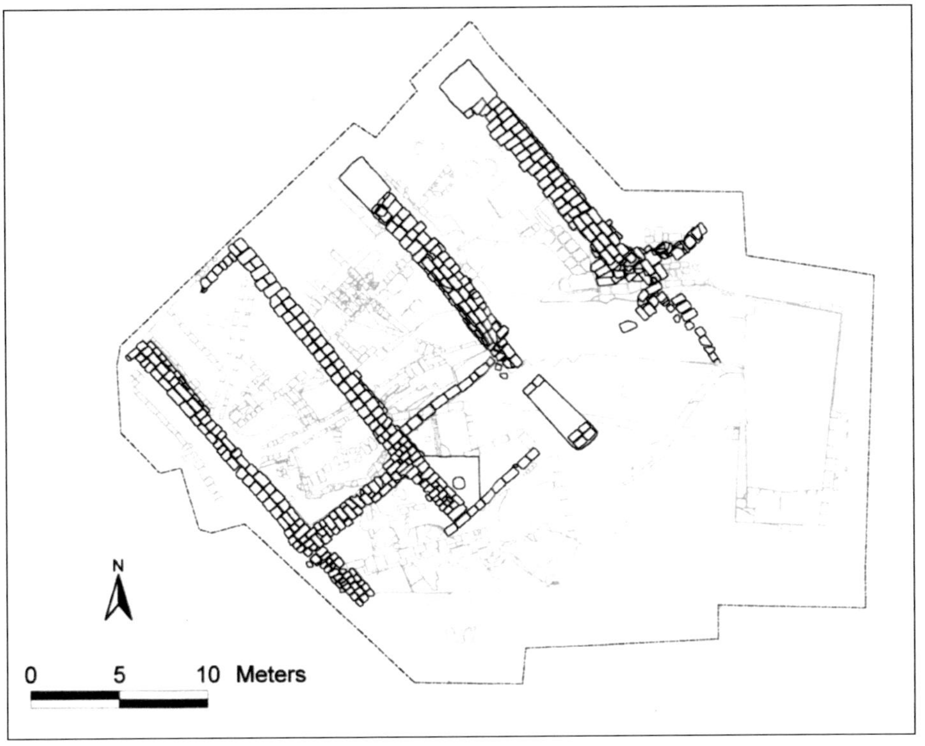

Figure 3c. Caere, Vigna Parrocchiale. Tuscan temple built on top of the Archaic residence (courtesy of the CNR – ISMA Archive).

clan, or better, *gens*) *haspna*, which implies that these weapons belonged to a private army of the *gens haspna*, which acted as a patron. This clan is otherwise unknown, but was clearly a powerful, maybe even ruling, family of Vetulonia around 500 BCE. This clan and its private army was most likely wiped out or cast out of Vetulonia around 470-450 BCE. Afterwards the helmets were ritually destroyed and buried with an obliteration rite on the *arx* of Vetulonia as symbols of the former regime.[9] This obliteration practice resembles the rituals in our case studies of Gabii and Murlo, Poggio Civitate.

Another case study could also be the *heroon* on the agora of the Greek city of Poseidonia. The 'tomb' and cult place of the mythical founder of Poseidonia was installed at the end of the sixth century BCE and became an important sanctuary at the centre of the *polis*. When Poseidonia became a Latin colony in 273 BCE, the Romans did not dare to destroy the sanctuary, but instead removed the *tumulus* and covered the tomb in a small enclosure with roof tiles (Greco, 2014). The meaning of this sanctuary was altered and diminished by covering and obliterating it, but not entirely removed from the public memory of Poseidonia.

9 The case is not as clear cut as it might seem here. According to Markus Egg (2017:p.173), the helmets were merely consecrations after a military victory in Vetulonia. Adriano Maggiani (1990:pp.48f.) originally emphasized an internal social upheaval in the city-state as the occasion for the deposition of the helmets, but recently seems to connect the depot to the Syracusean invasion of the mining area of Vetulonia (Maggiani, 2012:p.67).

Case Study	Chronology	Action
Rome, *regia*	Late 7th – end of 6th centuries BCE	Re-use, conservation
Gabii, *regia*	Second half 7th century – end 6th/beginning 5th centuries BCE	Obliteration, ritual burial
Murlo, Poggio Civitate	680/670 – 530 BCE	Obliteration, ritual burial
Caere, Vigna Parrocchiale	6th century – end 6th/beginning 5th centuries. BCE	Destruction, replacement

Other examples		
Deposit on the *arx* of Vetulonia	First half of the 5th century BCE	Obliteration, ritual burial
Poseidonia, *heroon*	Around 273 BCE	Obliteration, ritual burial
adfectatio regni in Rome	5th – early 4th centuries BCE	Destruction, replacement

Table1. Overview of case studies and 'cleansing practices' discussed here.

Further literary evidence: cases of adfectatio regni

Roman sources mention an interesting strategy during the early Roman Republic that might help us to better understand our case studies. In Roman historiography, there are at least three cases of *adfectores regni* – successful politicians or members of public life that aimed at tyrannical rule – in the early Republic: Sp. Cassius, 483 BCE; Sp. Maelius, 434 BCE; M. Manlius Capitolinus, 384 BCE (Smith, 2006; Torelli, 2017; both with references to the individual cases and the literary sources). All cases of *adfectatio regni* end with the condemnation and killing of the men that tried to rule as tyrants. Interesting for this topic is the fate of the houses (or better yet, the seats of power) of the three aspiring tyrants. Sp. Cassius' house was razed to the ground and replaced with a temple for Tellus.; the house of Sp. Maelius was surrogated by a public square named *Aequimelium*, while the house of M. Manlius Capitolinus made way for the temple of Iuno Moneta. In all cases, the houses of politicians that gained too much power and tried to become tyrants were destroyed and replaced with public monuments, often temples. This destructive practice seems to be a direct parallel to the case of Vigna Parrocchiale, where the palace was also replaced by a Tuscan temple.

Cleansing practices, curses and desecrations

The removal of persons or institutions from public space or public memory, as discussed here, was understood in antiquity less as a cleansing practice, but rather as a cursing or desecration in a religious sense. Curses by collectives are attested several times in antiquity (Graf, 2005:pp.250-253). In the Ionian *polis* Teos, threats to the democratic order were collectively cursed in the phase shortly after 479 BCE, that is, immediately after liberation from the Persians and during the domestic political danger of oligarchic upheavals. These curses were pronounced by the community at public festivals and cult celebrations and recorded on *stelae* (Graf, 2005:pp.250 f. no. 10-11). This is reminiscent of Roman actions against *adfectores regni*. Curses, exiles and death sentences (even in absentia) were apparently commonly combined against potential oligarchs and threats to public order (such as against Alkibiades in Athens). In this sense, the obliteration of the seats of such potential autocrats may also be interpreted as collective curses.

Another perspective on the obliteration of seats of power is religious profanation or desecration (Burkert, 2005). The sanctity, religious meaning and (sacred) protection of objects, buildings, sanctuaries and even cities could be withdrawn through rituals. In Roman religion this was done by *exauguratio* (in contrast to (*in-*)*auguratio*, *i.e.* consecration; Burkert, 2005:pp.271-273). During the conquest or siege of an enemy city, Romans were able to expel protective deities by *evocatio* and include them in their pantheon, as in the case of Juno during the conquest of Veii in 396 BCE (Liv. 5, 21). The permanent destruction or 'un-founding' of hostile places by ritual acts has been attested for the Romans (the most famous example being Carthage) and in the Old Testament (Edlund-Berry, 1994:pp.17f. with references; Burkert, 2005:pp.273f.). This, too, may be the cultural background for obliteration acts of seats of power.

Conclusions

The four case studies, as well as other archaeological evidence, have shown that notions of the *cleansing* of a political system in Central Italy of the seventh and sixth century BCE may include, but are not limited to:

1. the destruction of palaces, replaced by sanctuaries and places for public meetings (Caere, Vigna Parrocchiale; cases of *adfectatio regni* in Rome);
2. ritual obliterations and 'burials' of buildings under large tumuli and mounds (Gabii, *regia*; Murlo, Poggio Civitate; similarly, Vetulonia and Poseidonia);
3. conservation of seats of power for different 'reuses' (Rome, *regia*).

The case of the Roman *regia* might be similar to those of Vigna Parrocchiale in Caere and the *adfectores regni*, since here too a palace was turned into a sanctuary, although not by its destruction but rather by the building complex being reused.

Obliteration practices, as well as the destruction and replacement of palaces by sanctuaries, have in common the redefinition of public space (both functionally and in respect to public memory). This space is, in a sense, irreversibly taken away from private uses or ambitions. The place where a tumulus covers a palace or a votive deposit marks a ritual act (a curse?) can no longer be reused, no more than a place where a newly erected sanctuary has been erected on top of a former palace or seat of power. The two central aspects of the actions discussed here may be the rendering of a site unusable (or blocking its use), while simultaneously providing a reinterpretation of public space and public memory. The use of curses and desecrations show that all actions could be religiously charged. This is hardly surprising, since politics and religion in Central Italy in the Archaic and early Classical periods were not only linked, but inseparable, from each other.

Well-defined standard practices for what to do in the case of a regime change do not seem to have existed in Antiquity and they do not exist for modern societies either. When the city-states of Central Italy in the late sixth and early fifth century BCE experienced the collapse of autocratic regimes they naturally found different ways of 'cleansing' their new political systems of the monumental relics left by former rulers. In Antiquity, as in modern times, societies try to find ways to highlight an undesirable past; to preserve this past as a warning example, to bury it ritually or to destroy it. As in modernity, specific and diverse strategies were found for various situations both in order to cleanse the past from public spaces and to remove public memories from society.

Bibliography

Aigner-Foresti, L. (2003) *Die Etrusker und das frühe Rom*. Geschichte kompakt. Antike. Darmstadt, Wissenschaftliche Buchgesellschaft.

Aigner Foresti (2009) Forme di organizzazione politica nella *Caere* arcaica. In: Bruni, S. (ed.) *Etruria e Italia preromana. Studi in onore di Giovannangelo Camporeale. Vol. 1.* Studia Erudita 4. Pisa, Fabrizio Serra Editore, pp. 9-15.

Amann, P. (2017) Society, 580-450 BCE. In: Naso, A. (ed.) *Etruscology*. Boston, De Gruyter, pp. 985-999.

Becker, H. W. (2013) Political Systems and Law. In: Turfa, J. M. (ed.) *The Etruscan World.* Abingdon, Oxon, Routledge, pp. 351-372.

Bellelli, V. (2008) Per una storia del santuario della Vigna Parrocchiale a Cerveteri. In: Dupré Raventós, X., Ribichini, S. & Verger, S. (eds.) *Saturnia tellus. Definizioni dello spazio consacrato in ambiente etrusco, italico, fenicio-punico, iberico e celtico. Atti del convegno internazionale svoltosi a Roma dal 10 al 12 novembre 2004*. Rome, Consiglio Nazionale delle Richerche, pp. 319-333.

Bizzarri, C. (2013) Etruscan Town Planning and Related Structures. In: Turfa, J. M. (ed.) *The Etruscan World.* Abingdon, Oxon, Routledge, pp. 708-720.

Burkert, W. (2005) Profanation. Rituelle Massnahmen gegen Kulte und Götter. In: *Thesaurus Cultus et Rituum Antiquorum 3. Divination – Prayer – Veneration – Hikesia – Asylia – Oath – Malediction – Profanation – Magic Rituals and Addendum to vol. II Consecration.* Los Angeles, Getty Publications, pp. 271-281.

Carnabuci, E. (2012) *Regia. Nuovi dati archeologici dagli appunti inediti di Giacomo Boni.* Lexicon Topographicum Urbis Romae – Supplementum V, Rome, Edizioni Quasar.

Cifani, G. (2017a) Small, Medium or Extra-long? Prolegomena to any Future Metaphysics on the Reconstructions of the Temple of Jupiter Optimus Maximus Capitolinus. In: Lulof, P. S. & Smith, C. J. (eds.) *The Age of Tarquinius Superbus. Central Italy in the Late 6th Century. Proceedings of the Conference* 'The Age of Tarquinius Superbus, A Paradigm Shift?' *Rome, 7-9 November 2013*. Babesch. Annual Papers on Mediterranean Archaeology, Supplement 29. Leuven, Peeters, pp. 113-122.

Cifani, G. (2017b) Ein Blick zurück. Rom – eine „etruskische" Metropole der Königszeit. In: Badisches Landesmuseum Karlsruhe (ed.) *Die Etrusker. Weltkultur im antiken Italien. Sonderausstellung im Badischen Landesmuseum Karlsruhe vom 16. Dezember 2017 bis 1. Juni 2018*. Darmstadt, WBG, pp. 320-324.

Coarelli, F. (1983) *Il Foro Romano. I. Periodo arcaico*. Rome, Edizioni Quasar.

Colonna, G. (2010) Cerveteri. In: Bruni, S. (ed.) *Gli Etruschi delle città. Fonti, ricerche e scavi.* Milan, Silvana, pp. 182-193.

Cornell, T. J. (1995) *The Beginnings of Rome. Italy and Rome from the Bronze Age to the Punic Wars (c. 1000-264 BC)*. Routledge History of the Ancient World. London, Routledge.

Cristofani, M. (1987) Cerveteri. In: Nenci, G. & Vallet, G. (eds.) *Bibliografia topografica della colonizzazione greca in Italia e nelle isole tirreniche V*. Pisa, Scuola normale superiore, pp. 251-266.

Cristofani, M (ed.) (1990) *La grande Roma dei Tarquini. Catalogo della mostra. Roma, Palazzo delle Esposizioni. 12 giugno – 30 settembre 1990*. Rome, L'ERMA di Bretschneider.

Cristofani, M. (ed.) (1992) *Lo scarico arcaico della Vigna Parrocchiale 1*. Caere 3.1 Rome, Consiglio nazionale delle ricerche.

Cristofani, M. (ed.) (1993) *Lo scarico arcaico della Vigna Parrocchiale 2*. Caere 3.2 Rome, Consiglio nazionale delle ricerche.

Cristofani, M. (ed.) (2003) *Vigna Parrocchiale. Scavi 1983-1989. Il santuario, la "residenza" e l'edificio ellittico*. Caere 4. Rome, Consiglio nazionale delle ricerche.

de Grummond, N. T. (1997) Poggio Civitate. A Turning Point. *Etruscan Studies*, 4, 23-40.

de Grummond, N. T. & Pieraccini, L. C. (eds.) (2016) *Caere*. Austin, University of Texas Press.

Edlund-Berry, I. E. M. (1994) Ritual Destruction of Cities and Sanctuaries. The 'Un-Founding' of the Archaic Monumental Building at Poggio Civitate (Murlo). In: De Puma, R. D. & Small, J. P. (eds.) *Murlo and the Etruscans. Art and Society in Ancient Etruria*. Wisconsin Studies in Classics. Madison, The University of Wisconsin Press, pp. 16-28.

Egg, M. (2017) War and Weaponry. In: Naso, A. (ed.) *Etruscology*. Boston, De Gruyter, pp. 165-177.

Fabbri, M. (2015) A Seat of Power in *Latium Vetus*. The Archaic Building Complex on the *Arx* of Gabii. In: Kistler, E., Öhlinger, B., Mohr, M. & Hoernes, M. (eds.) *Sanctuaries and the Power of Consumption. Networking and the Formation of Elites in the Archaic Western Mediterranean World. Proceedings of the International Conference in Innsbruck, 20th-23rd March 2012*. Wiesbaden, Harassowitz, pp. 187-203.

Fabbri, M. (2017) La *regia* di *Gabii* nell'età dei Tarquini. In: Lulof, P. S. & Smith, C. J. (eds.) *The Age of Tarquinius Superbus. Central Italy in the Late 6th Century. Proceedings of the Conference* 'The Age of Tarquinius Superbus, A Paradigm Shift?' *Rome, 7-9 November 2013*. Babesch. Annual Papers on Mediterranean Archaeology, Supplement 29. Leuven, Peeters, pp. 225-239.

Fabbri, M. & Musco, S. (2016) Nuove ricerche sulle fortificazioni di Gabii. I tratti nord-orientale e settentrionale. In: Fontaine, P. & Helas, S. (eds.) *Fortificazioni arcaiche del Latium vetus e dell'Etruria meridionale (IX-VI sec. a. C.). Stratigrafia, cronologia e urbanizzazione. Atti delle Giornate di Studio, Roma, Academia Belgica, 19-20 settembre 2013*. Brussels, Belgisch Historisch Instituut te Rome, pp. 71-90.

Farrell, S. (2008) Firdos Square's Symbols: Then and Now. *The New York Times*. [Online] 11. December 2008. Available from: https://atwar.blogs.nytimes.com/2008/12/11/photographers-journal-firdos-squares-symbols-then-and-now/ [Accessed: unknown (eds.)].

Glinister, F. (2006) Kingship and Tyranny in Archaic Rome. In: Lewis, S. (ed.) *Ancient Tyranny*. Edinburgh, Edinburgh University Press, pp. 17-32.

Graf, F. (2005) Fluch und Verwünschung. In: *Thesaurus Cultus et Rituum Antiquorum 3. Divination – Prayer – Veneration – Hikesia – Asylia – Oath – Malediction – Profanation – Magic Rituals and Addendum to vol. II Consecration*. Los Angeles, Getty Publications, pp. 247-270.

Greco, E. (2014) *La 'tomba' del fondatore e le origini di Poseidonia*. Quaderni di Antichtità Pestane 3. Paestum, Pandemos.

Hopkins, J. N. (2012) The Capitoline Temple and the Effects of Monumentality on Roman Temple Design. In: Thomas, M. L. & Meyers, G. E. (eds.) *Monumentality in Etruscan and Early Roman Architecture. Ideology and Innovation*. Austin, University of Texas Press, pp. 111-138.

Hopkins, J. N. (2016) *The Genesis of Roman Architecture*. New Haven, Yale University Press.

Ion, E., Elias, L., Li, Y. & Mironova, J. (2014/2015) *Mapping Cultural Space Across Eurasia. Raining Lenins*. [Online] Available from: http://dighist.fas.harvard.edu/projects/eurasia/exhibits/show/raining_lenins [Accessed: unknown (eds.)]

Krämer, R. P. (2016) Trading Goods – Trading Gods. Greek Sanctuaries in the Mediterranean and their Role as emporia and 'Ports of Trade' (7th-6th Century BCE). *Distant Worlds Journal.* [Online] 1, 75-98. Available from: https://doi.org/10.11588/dwj.2016.1.30154 .

Maffei, A. & Nastasi, F. (eds.) (1990) *Caere e il suo territorio. Da Agylla a Centumcellae.* Rome, Istituto Poligrafico e Zecca dello Stato.

Maggiani, A. (1990) La situazione archeologica dell'Etruria settentrionale nel V sec. A. C. In: *Crise et transformation des sociétés archaïques de l'Italie antique au Ve siècle av. J.-C. Actes de la table ronde organisée par l'École française de Rome et l'Unité de recherches étrusco-italiques associée au CNRS (UA 1132). Rome 19-21 novembre 1987.* Collection de l'École française de Rome 137. Rome, École française de Rome, pp. 23-49.

Maggiani, A. (2000) Republican Political Forms. In: Torelli, M. (ed.) *The Etruscans.* Milan, Bompiani, pp. 227-241.

Maggiani, A. (2005) Da Veio a Vulci: le istituzioni politiche. In: Paoletti, O. (ed.) *Dinamiche di sviluppo delle città nell'Etruria meridionale. Veio, Caere, Tarquinia, Vulci, Atti del XXIII convegno di Studi Etruschi ed Italici. Roma – Veio – Cerveteri/Pyrgi – Tarquinia – Tuscania – Vulci – Viterbo. 1-6 ottobre 2001. Vol. 1.* Istituti nazionale di studi etruschi ed italici. Atti di convegni 23. Pisa, Istituti editoriali e poligrafici internazionali, pp. 61-69.

Maggiani, A. (2012) Das Helmdepot von Arce, Vetulonia. In: Meighörner, W. (ed.) *Waffen für die Götter. Krieger Trophäen Heiligtümer.* Innsbruck, Tiroler Landesmuseum, pp. 63-67.

Marcattili, F. (2005) Regia. In: *Thesaurus Cultus et Rituum Antiquorum 4. Cult Places. Representations of Cult Places.* Los Angeles, Getty Publications, pp. 308-311.

Menichetti, M. (2017) Die etruskische Stadt. Grundlagen von Gesellschaft und Verwaltung. In: Badisches Landesmuseum Karlsruhe (ed.) *Die Etrusker. Weltkultur im antiken Italien. Sonderausstellung im Badischen Landesmuseum Karlsruhe vom 16. Dezember 2017 bis 1. Juni 2018.* Darmstadt, WBG, pp. 173-177.

Moretti Sgubini, A. M. (ed.) (2001) *Veio, Cerveteri, Vulci. Città d'Etruria a confronto. Catalogo della Mostra. Roma, Museo Nazionale Etrusco di Villa Giulia. Villa Poniatowski. 1 ottobre – 30 dicembre 2001.* Rom, L'ERMA di Bretschneider, Ingegneria per la cultura.

Mura Sommella, A. (2009) Il Tempio di Giove Capitolino. Una nuova proposta di lettura. *Annali della fondazione per il museo Claudio Faina,* 16, 333-372.

Murgan, A. (2011) Regia. In: Steinmann, B., Nawracala, R. & Boss, M. (eds.) *Im Zentrum der Macht. Das Forum Romanum im Modell.* Erlangen-Nürnberg, Institut für Klassische Archäologie. Antikensammlung, Universität Erlangen-Nürnberg, pp. 66-71.

Pandolfini Angeletti, M. (1989) I dati epigrafici dallo scarico arcaico della Vigna Parrocchiale. In: Consiglio Nazionale delle Ricerche (ed.) *Miscellanea Ceretana 1.* Quaderni del centro di studio per l'archeologia etrusco-italica 17. Rom, Consiglio Nazionale delle Ricerche, pp. 69-83.

Potts, C. R. (2015) *Religious Architecture in Latium and Etruria, c. 900-500 BC.* Oxford Monographs on Classical Archaeology. Oxford, Oxford University Press.

Prayon, F. (2010) Frühetruskische Hausarchitektur. Bemerkungen zum Forschungsstand. In: Bentz, M. & Reusser, C. (eds.) *Etruskisch-italische und römisch-republikanische Häuser.* Studien zur antiken Stadt 9. Wiesbaden, Dr. Ludwig Reichert Verlag, pp. 9-28.

Pucci, E. (2016) Elmi dell'Arce di Vetulonia. Castiglione della Pescaia (GR). Il restauro di un gruppo di elmi di tipo Negau-Vetulonia. *Notiziario della Soprintendenza per i Beni Archeologici della Toscana. Saggi, 11/2015,* 457-463.

Sassatelli, G. (2017) '... die Heimat von Städten'. Die Entwicklung der etruskischen Stadtstaaten. In: Badisches Landesmuseum Karlsruhe (ed.) *Die Etrusker. Weltkultur im antiken Italien. Sonderausstellung im Badischen Landesmuseum Karlsruhe vom 16. Dezember 2017 bis 1. Juni 2018*. Darmstadt, WBG, pp. 162-168.

Scott, R. T. (1999) Regia. In: Steinby, E. M. (ed.) *Lexicon Topographicum Urbis Romae. Volume Quarto P-S*. Rome, Edizioni Quasar, pp. 189-192.

Smith, C. J. (1996) *Early Rome and Latium. Economy and Society c. 1000 to 500 BC*. Oxford, Oxford University Press.

Smith, C. (2006) Adfectatio regni *in the Roman Republic*. In: Lewis, S. (ed.) *Ancient Tyranny*. Edinburgh, Edinburgh University Press, pp. 49-64.

Tagliamonte, G. (2017) Political Organization and Magistrates. In: Naso, A. (ed.) *Etruscology*. Boston, De Gruyter, pp. 121-141.

Torelli, M. (1990) La società etrusca della crisi. Quali trasformazioni sociali?. In: *Crise et transformation des sociétés archaïques de l'Italie antique au Ve siècle av. J.-C. Actes de la table ronde organisée par l'École française de Rome et l'Unité de recherches étrusco*-italiques associée au CNRS (UA 1132). Rome 19-21 November 1987. Collection de l'École française de Rome 137. Rome, Ècole française de Rome, pp. 189-198.

Torelli, M. (2000) Le *regiae* etrusche e laziali tra Orientalizzante e Arcaismo. In: Bartoloni, G., Delpino, F., Govi, C. M. & Sassatelli, G. (eds.) *Principi etruschi tra Mediterraneo ed Europa*. Catalogo della mostra Bologna. Museo Civico Archeologico. 1 ottobre 2000 – 1 aprile 2001. Venice, Marsilio, pp. 67-78.

Torelli, M. (2017) Templi sopra case. Archeologia dell'*adfectatio regni*. In: Lulof, P. S. & Smith, C. J. (eds.) *The Age of Tarquinius Superbus. Central Italy in the Late 6th Century. Proceedings of the Conference* 'The Age of Tarquinius Superbus, A Paradigm Shift?' *Rome, 7-9 November 2013*. Babesch. Annual Papers on Mediterranean Archaeology, Supplement 29. Leuven, Peeters, pp. 39-48.

Tuck, A. (2014) Manufacturing at Poggio Civitate. Elite Consumption and Social Organization in the Etruscan Seventh Century. *Etruscan Studies*, 17 (2), 121-139.

Tuck, A. (2015) Poggio Civitate. 2014 Field Report. *Etruscan Studies*, 18 (1), 28-39.

Tuck, A., Brunk, J., Huntsman, T. & Tallman, H. (2010) An Archaic Period Well from Poggio Civitate. Evidence for Broader Final Destruction. *Etruscan Studies*, 13, 93-104.

Tuck, A., Glennie, A., Kreindler, K., O'Donoghue, E. & Polisini, C. (2016) 2015 Excavations at Poggio Civitate and Vescovado di Murlo (Provincia di Siena). *Etruscan Studies*, 19 (1), 87-148.

Tuck, A. & Wallace, R. (2013) *First Words. The Archaeology of Language at Poggio Civitate*. Hanover, Pennsylvania, Sheridan Press.

Tuck, A., Wallace, R., Whitcher Kansa, S. & Horvitz, C. (2015) *Vinum. Poggio Civitate and the Goddess of Wine*. Hanover, Pennsylvania, Sheridan Press.

van't Lindenhout, E. (2017) Looking for a Royal Residence. Architecture and Social Complexity in Archaic Rome. In: Lulof, P. S. & Smith, C. J. (eds.) *The Age of Tarquinius Superbus. Central Italy in the Late 6th Century. Proceedings of the Conference* 'The Age of Tarquinius Superbus, A Paradigm Shift?' *Rome, 7-9 November 2013*. Babesch. Annual Papers on Mediterranean Archaeology, Supplement 29. Leuven, Peeters, pp. 87-95.

Cleaning up the Past

Mareike Späth

Collective memory is formed through a process of selection.[1] Societies create their past in the present by deciding to commemorate and to forget. Memory entrepreneurs[2] remove tangible and intangible remnants that bear witness to an unwanted past. In short: we clean up because the past is a messy affair. This is true especially of moments in time when power structures change and values shift. Such cleaning practices are diverse: the toppling of monuments, the burning of archives, the rushed disposal of corpses, or the eradication of remembrance days from the commemorative calendar, to list just a few.

How is a difficult or unwanted past eliminated through cleaning practices? How do societies decide which parts and pieces of the past are worth keeping and what should be discarded? Who is commissioned with this task? Why is it so difficult or even impossible to dispose of the past? This paper looks at ways of dealing with the past through cleaning practices, particularly using the example of different monuments. I will look at a specific form of cleaning: cleaning up in the sense of putting away and clearing or rearranging monuments that have changed their value in the eye of the beholder.

I depart from the idea that a condition of cleanness is established through the act of putting away "matter out of place" (Douglas, 1966:p.35). Things out of place, like monuments that represent a former political power, are often perceived as something that is staining the monumental landscape and simoultaneously contaminating and polluting collective memory. So monument landscapes as ensembles of political and cultural symbols are adapted to changed conditions by cleaning up, that is by throwing out, tearing down or repositioning monuments. Thus, the monumental landscape adapts to the new conditions and is liberated from old stains. A clean condition is re-established. The treatment of monuments of course reveals the ambiguities immanent in global entangled histories and sheds light on the question of ownership of history. Drawing from existing anthropological and historical literature, artworks, and my own empirical research in Madagascar, I argue that looking at the treatment of visible

1 I thank Isabel Bredenbröker, Hans Peter Hahn and Kathrin Knodel for sharing their thoughts and commenting on earlier versions of this essay.

2 For a discussion of the 'entrepreneur' in memory making projects see N'Guessan and Späth, 2016:p.29f.

in: Bredenbröker, I., Hanzen, C. & Kotzur, F (eds) 2020: *Cleaning and Value. Interdisciplinary Investigations*, Sidestone Press (Leiden), pp. 213-228.

representations of history as an act of cleaning up will reveal how certain aspects of the past are held as meaningful (or not) for the present condition of a given society.

Monuments as Objects of History

In line with Pierre Nora (1989), I understand monuments as media that communicate a certain *lieux de mémoire* to a community of shared memory. Their construction is an act of attributing value to a person, an event or an ideology. The will to remember, combined with a specific interpretation of the given historical events or persons, is inscribed into a monument which communicates this perspective to a public audience. The monuments that a society builds visualize 'who they {the society} believe they are' (Zerubavel, 2003:p.319).

But politics and ideologies change. After a change of government, the individuals or events represented in the monument landscape often fall from grace. Their monuments are now traces of former political powers. Following Mary Douglas' seminal definition of dirt as 'matter out of place' (1966:p.35), these monuments are matters physically still in place, but at the same time out of place, since they belong to a symbolic order that has ceased to exist. Ideologically outdated, they offend against the new order. In other words, they are 'difficult heritage', the heritage people wish they didn't have (Macdonald, 2009). How do individuals, governments and societies deal with such monuments? Annelise Finney argues that bureaucracies tend to create monuments as political propaganda, which are 'vulnerable to future deletion and abandonment' (2015). Suddenly, monuments are no longer ornamentation but an eyesore calling for removal. Finney describes this as 'the cycle of impermanence' (2015).

Collective history is never consensual, and objects defy from definite and unchangeable meaning. Consequently, the issue of monuments is complex. Within a society, heterogeneous views of the past coexist or clash, and a certain monument can have different meanings, depending on the interpretation of the onlooker. Given these simultaneities, the process of handling monuments is one of negotiating the value of memories and heritage. The construction of a monument is often, to a large part, a bureaucratic act and it is usually representatives of the state who controls the public landscape and the monuments within public space. The same is true for erasing monuments. However, civil heritage entrepreneurs or actors of the political opposition do also initiate the building, transferral or demolition of monuments. They can deal with existing monuments in creative ways to express alternative views. Hans-Rudolf Meier considers monuments 'sand in the gears' and 'stumbling stones' (2013:p.18, my translation). They have the potential to remind people of possible alternatives and to document a different, outdated or unconventional interpretation of the past. In other words, whether a monument is valuable, or trash is a question of perspective. Different answers to this question coexist within a society and change in the course of time. Negotiating and handling monuments, keeping, maintaining or discarding them, is therefore part of negotiating the value of the past. from a present point of view. In the long run, the result of such negotiations shows whose voices in the cacophony of memory entrepreneurs are being heard and being cared for by those who are in a position of power and therefore orchestrate the public space we all inhabit.

Figure 1. A crowd symbolically besmirching the First Malagasy Republic in Antananarivo, Madagascar (Photo: Mareike Späth).

Cleaning and Staining

The practice of building a monument or a statue is an act of honouring and attributing positive value. The maintenance and preservation of monuments by means of cleaning their surfaces, scraping off dirt and moss and renewing colors or inscriptions is a way of expressing respect to those the monument is dedicated to. Often such activities are included as parts of commemorative rituals. Whereas the maintenance of (larger) monuments is often professionalized and budgeted by the state, the cleaning and caring for monuments by individuals can also be a community activity.[3]

In contrast, deteriorating or dirty monuments are often perceived as disfiguring the physical and symbolical landscape. Unruly behaviour towards a monument, too, can be interpreted as an act of devaluation.[4] Neglect and lack of maintenance does harm to monuments, but even more so deliberate soiling, for example with paint or signs of protest. Protesters express their discontent with a political ideology by 'rituals of desecration' like besmirching, staining or damaging. People turn against monuments during revolutions or in the immediate aftermath of political change. These are symbolic and visible acts of rupture and liberation. Cleaning stains off the violated monuments, as a contrary kind

3 Think for example of the ritualized cleaning of tombstones before Veterans Day.

4 The state usually sanctions any spoiling or damaging of monuments. In Germany for example § 303 StGB classifies the spoiling or damaging of monuments as *Gemeinschädliche Sachbeschädigung* (damage to property), offenders can be punished with a fine or imprisonment up to three years.

of activity, means restoring value and dignity to a monument with its associated subject matter that is being commemorated. Deliberate staining of monuments and subsequent cleaning is a discursive practice, a comment on the subject matter of collective memory in its public, material manifestation.

While conducting research on commemorative practices in Madagascar, I attended a performance that was part of the jubilee celebrating 50 years of Malagasy independence from France. A senior Malagasy lady pointed to a scene at the side of the stage: Onlookers were standing and sitting on a monument to get a better view, some were leaning against it to relax from hours of standing, other were sitting underneath, seeking shade.

The monument in question is dedicated to the First Malagasy Republic (1958-1960), proclaimed in 1958 within the newly created French Community. Today, this first postcolonial government is criticised by the lady next to me and many other Malagasy people for its continued close collaboration with France. She drew my attention to the way the people 'used' the monument and explained their behaviour as vilification. She was obviously pleased to witness what she interpreted as public demonstration of disrespect towards a former Malagasy government that tolerated or even profited from French (post)colonial influence in Madagascar. In her eyes, the display of inappropriate behaviour like sitting on a monument showed a public demonstration of disrespect towards the Franco-Malagasy political alliance and the amnesia of colonial violence which the monument – in her eyes – represents.

Removal

A study for the European Union that looked at the removal of historical traces related to dictatorships and humanitarian crimes within Europe found that 'with few exceptions (*i.e.* Spain), EU member states accomplished relatively quickly the removal of all symbols related to the repressive past.' (Montero, 2010:p.294). The elimination of monuments can be a 'revaluation or devaluation of objects (Hahn, 2016:p.12), a 'positive effort to organize the environment', (Douglas, 1966:p.2) and may work through the logic of manufacturing a 'systematic adjustment of pre-history' (Hinz, 1993:p.304).

A strong and effective way of dealing with unwanted monuments is to remove them. When British and French colonial government took over the German overseas territories at the end of World War I, they immediately removed all Bismarck statues and similar colonial monuments (Speitkamp, 2000:p.178). On 13 Mai 1946, only 5 days after the official end of World War II, the Allied Control Council issued Directive 30, the command to remove all German military and Nazi monuments and museums.[5] The Council made the end of the Nazi regime visible and prevented continued public veneration of Nazi monuments (and ideology). This was just recently applauded by American commentators when in 2017 the question of how to deal with confederate monuments was heatedly debated in relation to the 'Rhodes Must Fall' movement at the University of Cape Town that started Fallism – the call for bringing down monuments and other symbols of white supremacy – worldwide.[6]

5 *Befehl Nr. 30* on the *Beseitigung deutscher militärischer und nazistischer Denkmäler und Museen*. Archiv Frankfurt (Oder), Bestandsabteilung II, Nr. 711 (Targiel, 2002:p.36). For a translation and full text of the directive refer to Dessem, 2017.

6 See e.g. Zeitz, 2017.

When the Ghanaian government under president Kwame Nkrumah was overthrown in February 1966, one of the first measures of the new military government was to ban all images of the former president from the public sphere. Newspapers reported that an angry mob attacked, beheaded, and pulled down the statue of Nkwame Nkrumah in front of Parliament House. A press photograph showed a group of children standing around the decapitated sculpture. This press coverage created an image of toppling Nkrumah as public act by the people, even children. However, as Carola Lentz convincingly argues, the statue was professionally brought down and 'the alleged spontaneity of the destruction [...] was a propaganda ploy of the coup makers who wanted to convince the world of the unpopularity of his regime' (Lentz, 2016; see also Lentz, 2017).

Similarly, videos of the statue of Saddam Hussein, dangling in mid-air in Firdos Square, Baghdad on April 9, 2003, were circulated widely via global media. Sean Aday and his colleagues argue that the toppling of Saddam Hussein's statue was largely staged by the US military instead of being a large spontaneous public event attended and cheered by masses, as international media coverage wanted the world believe. These media images, according to Aday, were supposed to signal the victorious end of the war. "Whereas battle stories imply a war is going on, statues falling – especially when placed in the context of truly climactic images from recent history – imply the war is over" (Aday, Cluverius & Livingston, 2005). This massage, Aday shows, was directed towards an American audience awaiting the end of the war.[7]

In November 1989, in the course of the political unification of the two Germanies, people in Berlin were 'attacking the material integrity of the border' (Klausmeier, 2013:p.158) between East and West. Although the Berlin Wall was not built for the purpose of being a monument but as a very real border, it had become a symbol of separation of the Germanies. So-called *Mauerspechte* were then tearing down the Berlin Wall bit by bit. Pictures document how a political decision was implemented physically by taking a symbol of division apart. Today, the consumption of the wall is remembered as a collective and symbolic act of liberation. It is not remembered as the execution of a directive or as a revolutionary act but as the appropriation of a state-dominated past by the liberated public including Germans from West and East, young and old. And while it marked a reaction to the German Democratic Republic politically collapsing, it was also an active part of this collapse, contributing to the event in a significant way and making it real by allowing people to cross the border and to deconstruct the fortification that previously separated states and people.

But political changes do not necessarily have to come with changes in the *monumentscape*. After Ukraine gained independence and the Soviet Union collapsed in 1991, most Lenin statues there were left in place. The new Ukrainian government ignored these remnants of the communist past. It was not until much later that the new president Yushchenko called for a 'removal of monuments of the totalitarian past' (cited in Mühling, 2011). This directive was vague enough to be ignored by Ukrainian bureaucracy. But the cleaning of the present from selected traces of the past is by no means monopolized by

7 See Fisher, 2011 for the argument that the media created a false impression of what happened on Firdos Square on April 9, 2003, and Speitkamp, 1997 and Hinz, 1993 for several well documented case studies which underline that the toppling of monuments is frequently a staged spectacle and only rarely a spontaneous outbreak of vandalism.

state institutions alone. In 2009, activists started scraping off the nose of the larger than life Lenin statue in Kiev (Mühling, 2011). The state in turn refurbished the statue. During protests against the new President Viktor Yanukovych's rapprochement towards Russia between December 2013 and February 2014, over a hundred Lenin statues and Soviet icons across Ukraine were vandalised or destroyed. Kiev's Lenin was brutally toppled and dismembered. 'Some took hammers to the fallen statue in an attempt to smash it to pieces,' *The Guardian* reported on December 8, 2013.[8] Since the Lenin monuments had not (all) been cleared after the fall of the communist regime, they were now targeted by protesters who directed their anger about current politics against the old soviet monuments.

But once a monument has fallen out of grace, been disfigured, taken apart or destroyed, what happens with its material body or the parts that remain? How to get rid of these unwanted reminders of the past?

Discarding and Storing

Cleaning up often means throwing things away and making sure that what is thrown away will not come back. According to Schofield, in Post-Nazi Germany '[o]fficials went so far as to pulverize the bricks [of Nazi monuments] and throw the remains into the North Sea' (2015). This is one way of ensuring complete disappearance. Radical elimination guarantees that no person who might still hold some values for the object in question will ever again put statues in a place of worship, on a pedestal or in a museum.

However, history's cleaning personnel regularly refrain from doing away completely with what was once valuable, and the throw-outs of history are rather stored away in the attics and cellars of (cultural) memory. Apparently, toppled monuments sometimes resist being thrown away. A obstinacy or permanence emanating from former glory is inherent to leftovers such as stones, bricks, dust, and metal (Speitkamp, 1997; Hahn, 2016:p.12). As a result, remainders of this past-to-be-forgotten often resurface and induce a (re) remembering of what had been buried in oblivion.

Unlike in post-soviet countries where the old regimes together with the USSR dissolved and ceased to exist, in the postcolonial context the former colonising regimes continue to exist within the boundaries of their European territories. In most cases, the parting colonial regimes took along their colonial monuments.[9] Where this was the case, cleaning up after themselves by removing the monuments off their former colonial territory can be read as an attempt by the colonising nations to keep and control their colonial monuments and with these the narrative of colonial grandeur.

Malagasy independence from France was officially declared in 1960. However, the first postcolonial government maintained a close relationship with France. Only in 1972, after a revolution led by the socialist movement which resulted in the declaration of the so-called Second Independence, the postcolonial Malagasy state began to remove stains of French colonial rule from its bureaucracy, its collective memory, and urban architecture. Sensing the threat for colonial monuments, the French ambassador to Madagascar, Maurice Delaunay, reported to the French government that the massive larger-than-live equestrian

---

8 Walker, S. (2013) Ukraine protesters topple Lenin statue in Kiev. *The Guardian*. [Online] 8th December 2013. Available from: https://www.theguardian.com/world/2013/dec/08/ukraine-opposition-viktor-yanukovych-european-integration [Accessed 24th March 2018]. The article includes a video of the act of demolition.

9 For Germany see Speitkamp, 2000:pp.178, 182f.

statue of General Gallieni, who according to the monument's inscription conquered the island in 1898 as heroic 'pacifier and organiser of Madagascar', had been taken down and safely stored away in French custody. The statue found an interim accommodation in the courtyard of a French military base. One year later, French military presence in Madagascar finally came to an end. When the armed forces withdrew, they amassed all signs of the colonial past and took them to France. Maréchal Gallieni's statue found a new spot to tell his story of out-of-place heroism at Place d'armes colonel Cazeilles at camp Lecocq in Fréjus, another French military camp, albeit without its impressive base.[10] Here, apparently, past military grandeur is still valued. This perspective, reflected in history books, tells a French story of saving valuables of a glorified past from Malagasy revenge.

As the example of French monuments in Madagascar shows, the cleaning up of statues by the French was not a service of goodwill to the formation of a French-free national narrative by newly independent Malagasy state. Instead, the French wanted to prevent the use of their valued symbols for the reformation of a new Madagaskar, which would have meant a possible destruction of their monuments. By holding on to them, they held on to their power and the ability to narrate the story of the colonial encounter and French grandeur. By bringing monuments back and preserving them, these could be repurposed on French home grounds.

But asking people in Madagascar about the destiny of statue, I have been told a popular, more patriotic version of the clearing of Gallieni from the surface of the Malagasy capital. 'Enraged Malagasy have torn down the statue to revenge the cruelties Gallieni had done to the Malagasy people', I was told repeatedly. This Malagasy version of what happened presents the process of cleaning as appropriating history by the Malagasy people and reclaiming power. The Malagasy historian Lucille Rabearimanana calls the clearing of Gallieni from the public space of Antananarivo by the French a 'manifestation of a politics that tends to eradicate the traces of colonial history' (my translation). In her perspective, the dismantling does not mean cleaning the colonial presence from Malagasy history, but is instead more an act of cleansing French collective memory from the stains of colonial debt and a distortion of historical facts (Pigenet & Tartakowsky, 2006:p.90).

Another French attic, or rather backyard, of colonial history is the Jardin d'agronomie tropicale in Vincennes at the eastern outskirts of Paris. The park was a place for promoting the idea of a colonial empire, for researching colonial plants and people, and bringing the overseas territories to an urban French audience. The Parisians once curiously pilgrimaged to Vincennes to visit the colonial exhibition in 1904 and look at colonial subjects on display. This also made it a suitable spot for storing away the remains of colonial representation after the empire had collapsed. Wandering in this park in January 2017, I came across the *Monument à la gloire de l'expansion coloniale française "Monument in honor of French colonial expansion"*, now scattered in pieces, covered with rotting leaves and overgrown with moss.

Built in 1913 by Baptiste Belloc, who was at that time official sculptor at the ministry of the colonies, the monument was built to 'materialize the beauty and the utility of the colonizing project' (cited in Aldrich, 2006:p.5). The monument toured

10 see Ranaivoson, 2005:p.79; Späth, 2015:pp.95-111; Oberlé, 1976 and Delauney, 2004:p.48. Abweichend Galibert, 2011:p.215, FN 96. Today, the space that the Gallieni momument once occupied is the site of a monument commemorating the martyrs of the anticolonial uprising in 1947.

Figure 2. Pieces of the *Monument à la gloire de l'expansion coloniale française* at the Jardin d'agronomie tropical (Photo: Mareike Späth).

around Paris before coming to the park. It figuratively displayed the metropolis and the overseas territories as a group of female figures: a naked Africa, Arabia with a turban, and seductive Asia, all looking up to France 'with devotion' as described in *La Dépêche coloniale illustrée* (cited in Aldrich, 2006:p.6, my translation). In 1961, the monument had become 'quite compromising' (Aldrich, 2006:p.6.) and was scattered across the park, fragmented into several pieces. The five pieces were later regrouped in one place and today the damaged fragments show 'the dramatic metamorphosis of colonial France' (Aldrich, 2006:p.7). The park has become a backyard where France stores the disquieting remains of her colonial representation. The park's website, however, considers it 'the most romantic park in Paris' precisely because of the many ruins of what is here obviously considered an unproblematic past.[11]

Not only the French struggle with the task of storing their colonial past out of sight. The adventurous biography of Carl Peters' statue demonstrates the ups and downs of a memorial. Its story was written by Peters' admirers and critics alike. Today, Carl Peters is one of the most controversial actors of German colonialism, known for his outstanding brutality. A statue of Peters was built by Karl Möbius in 1914 and sent to Dar es Salaam. Germany lost the war and its colonies before the statue was erected. Some German traditionalists negotiated that the colonial monument should be sent back from Dar es Salaam to Germany in 1919 (Schmid, 2014). Like the monument of his fellow colonial hero Wissmann, Carl Peters' statue was meant to be erected in Hamburg harbour as a reminder of Germany's ongoing colonial claims. However, because of the controversies concerning his person, he was banished from public and relegated to the basement of a mariner's house (Speitkamp, 2000:p.183). On the initiative of Franz Xaver Ritter von Epp, a protagonist of colonial revisionism

11 Office de Tourisme Vincennes. (n.d.) *Le Jardin d'Agronomie Tropicale*. [Online] Available from: http://www.vincennes-tourisme.fr/Decouvrir/Bois-de-Vincennes-et-ses-alentours/Le-Jardin-d-Agronomie-Tropicale (Accessed 15th May 2017).

and member of the NSDAP, the statue was erected on Helgoland in 1931.[12] During World War II, the statue was brought down and dismembered into pieces.[13] In the late 1940s, the body of the statue resurfaced in a wasteyard in Bremen, from where it was salvaged, only to be locked away again into a basement in Hamburg-Pinneberg. After the demolition of the building, workers sold the monument's individual pieces to a scrap merchant, who took them to be the remainders of a Göring statue and was probably hoping to sell them on the black market. However, he was sentenced to prison and the pieces were secured when the trade was busted (Schmid, 2014). In 1966, members of a colonialist group arranged a public display of the bust in front of the youth hostel in Helgoland. In 1989, alarmed by an article written by journalist Eckhard Groth that addressed Carl Peters' colonial crimes and his posthumous career as hero of the far right, local bureaucrats decided to finally 'put the bust away' (Schmid, 2014). But the monument was still not discarded. Today, it resides at a museum in Helgoland, where it is indeed 'put away' rather than displayed. It is displayed without any explanatory tag in the yard of the museum. Several websites and lay historians have written about the monument's lively past and present. They make frequent use of words related to dirt and cleaning. Images of the bust are frequently tagged as showing 'junk' and the 'rubbish' of history, combined with the demand that it should be 'tidied up' from public space once and for all.[14]

So how to keep control over the remnants of toppled or outdated and therefore deconstructed monuments? In Hungary, the dilemma of where to keep all the Soviet monuments fallen out of time was solved by establishing Szoborpark or Memento Park near Budapest. 42 giant monuments of communist dictatorship, formerly distributed all over the national territory, were assembled and recycled as a tourist attraction and education centre. The guided tour I participated in conveyed in some ironizing tone alternative histories of how these monuments were secretly mocked by Hungarians during their 'active time'. The subtext of the tour disclosed the seducing function of the giant monuments during the time of the Soviet Union.

Likewise, the toppled statue of Kwame Nkrumah is now exhibited in its disfigured state at the Kwame Nkrumah Memorial Park in Accra, Ghana. In this case, the narrative told by the operator of the park dismisses the act of breaking and toppling the statute as vandalizing it. The inscription honours the man who recovered the pieces as a 'patriotic citizen'. Reassembling, but not restoring, the monument therefore discredits the statue's aggressors, not the former head of state whose memento has politically been restored in the meantime.

Reusing and Recycling

A common way to deal with monuments that do not serve their original purpose anymore is reuse. The famous elephant statue in Bremen has turned from a colonial monument into a reminder of colonial violence and today stands as a symbol of

12 This site was chosen because Peters was involved with the Helgoland-Zanzibar Treaty of 1890, when the German Empire gained the island of Helgoland. The Nazi regime in general showed great ambition in restoring colonialism monuments to underline their political ambitions and territorial expansion.

13 Information on how and why are contradicting. Either it was damaged during an air raid or it was designated to melted down and recycled into ammunition.

14 See e.g. https://halunder.wordpress.com/tag/geschichte/ [Accessed 23rd June 2017].

Figure 3a-b. Discarded Soviet monuments at the Memento Park near Budapest (Photo: Mareike Späth).

Figure 4. Kwame Nkrumah's statue displayed in two pieces at the Kwame Nkrumah Memorial Park in Accra, Ghana. (Photo: Mareike Späth).

responsibilities that arise out of past colonial atrocities, as the new inscription reads. In other examples, statutes have been kept, but their names and inscriptions were altered to meet new mnemonic demands. Former inscriptions of 'old' statues are sometimes scraped off and replaced or overwritten with new ones, so as to tell new heroic narratives more suitable to changed histories. The statues continue to exist, albeit with a new identity (for an example from Egypt see Hinz, 1993:p.304).

Removed statues and monuments are often recycled and reused in creative ways. With the intention to confront past and present and to bridge the gap left behind by toppling the Lenin statues, German artist Rudolf Herz tied three busts formerly belonging to a Lenin monument in Dresden onto the back of a truck and toured 8000 kilometres through Europe in 2004. The artist staged this 'theatrical vivcation' (*theatralische Verlebendigung*) to address and to question the interrelation of memory, future and utopia (Herz, 2009). Finally, the bits and pieces of the Berlin Wall have not dissolved into dust. Quite the contrary, they continue to become more valuable. Segments of the Berlin Wall still in place have today become monuments, and others have been transported to cities all over Germany where they figure now as public monuments remembering the unification. Today, even the tiniest (and often fake) fragment of the Berlin Wall is turned into a precious souvenir and traded to tourists and nostalgic Germans. Disjointed, the symbol of division has been recycled as a symbol of reunification. In bits and pieces, the public wall has entered private space where it now decorates shelves and desks.

All these examples illustrate the 'stubbornness' of monuments, how they resist being thrown away. The monument biographies I have assembled show that although monuments may be cleared away and discarded by someone at some point in time, they rarely end up as trash or dirt. The materiality of monuments maintains its saturation with historical meaning and usually continues to be treated as matter of special status.

Purifying

The complete destruction of monuments may eventually be understood as an acknowledgement of their potential future power. But simply erasing tangible evidence does not undo history. The physical tearing down and disposing of monuments – the visible and factual clearing of the *monumentscape* – neither ensures a political nor cultural transformation, let alone actual forgetting. It can only produce a superficial amnesia. Once we take a closer look, we realize that beneath the seemingly clean surface, the past still sticks to the collective memory. It needs some more profound practice than dusting off, clearing, and sweeping to get rid of all the layers of mnemonic dirt that as accumulated in the past of a society.

The Mexican artist Isa Carrillo has produced an installation-performance for the purification of monuments. Let's go back to Kiev and revisit the site of the former Lenin monument. The statue itself has been torn down by protesters. The clearing, however, was only half-completed: a giant pedestal still marks the location of the former statue. Since then, numerous interventions have taken over the task of symbolically cleaning the site. Carillo has created a work which was part of a project called 'Social Contract', curated by Kateryna Filyuk. As part of this series of artistic interventions, Carillo's work used the abandoned pedestal of the toppled Lenin statue in Kiev to fuel a public discussion on the status and functions of commemorative objects in urban public space. Carillo proposed 'to neutralise' the ideological tension around the vacant pedestal by 'taking over' the monument by living plants' (Izolyatsia, 2017).[15] Negative energy at the site was supposed to be counteracted with rosemary and mint plants, using the healing powers of nature to 'purify the site' which is associated with the atrocities of former political regimes. The artist also invited citizens to participate and use the intervention to seek inner harmony. At the finissage of the installation, the herbs were harvested and transformed into purifying mint tea. The tea was given to passers-by and they were invited to participate in the purification of collective memory by means of their own bodies. This bodily purifying, the intervention suggests, runs far deeper than superficial (and unfinished) removals of monuments. It is an effort to clean up traces of past experiences that could not have been removed with the statue but still remain within the social body as well as within the bodies of individuals.

Dealing with Cleanliness

In ancient Rome, *damnatio memoriae*, the posthumous condemnation of the memory of a person, was practiced to punish traitors or those who brought discredit to the Empire. But it was difficult to implement this ban to a full extend. In fact, these measures did by no means lead to forgetting the person in question. Rather, the memory of a person was consciously kept alive through this process. Hedrick thus speaks of a 'memory of forgetting' (Hedrick, 2000). The act of removing a monument always leaves a noticeable gap and does not undo the past but may instead highlight it.

Similarly, refraining from constructing monuments can be considered a stain. In May 2014, the German/Ghanaian artist Philip Metz was invited by the *Kulturbüro Karlsruhe* to create a piece of art dealing with the theme of German colonialism or rather with the theme of the so-called colonial omission. Alongside other artists, Metz was asked to return the *Mémoires Perdues* (lost memories) to the present, and

---

15 Izolyatsia. (2017) *Ritual of Self-Nature: parallel events*. [Online] Available from: https://izolyatsia.org/en/project/social-contract-2017/public-programme/ [Accessed 23rd March 2018].

(re)introduce aspects of the past into historiography that had formerly been left untold.[16] Francois & Schulze's (2001) publication on *Deutsche Erinnerungorte* (German locations of remembrance, the pendant to the seminal French *Lieux de Mémoire* by Pierre Nora) neither addresses colonialism nor any colonial monuments. This void is the result of efficient and deliberate cleansing. Researching into the colonial memory of Germany, the artist found nothing but a 'blind spot' in German historiography. He found himself facing the odd task of creating a monument in the face of a past that looks suspiciously clean. Consequently, he considered it necessary to refurbish (or repollute, if you will) the *monumentscape* through interventions. Metz created a work entitled *Le héros invisible* (The Invisible Hero): A pedestal without statue. He alludes to a monument that has been cleared away. The inscription reads 'In honor to the Cameroonian soldiers fallen for the German fatherland – 1914-1918.'[17] He hereby reminds the public of those African soldiers who were forced to fight for the colonial powers in World War I but who have largely been forgotten or neglected by postwar European historiography and who have never had their monument. This work is part of a series named *Adler Afrika* which deals with German history on the African continent and points out that it takes a political intention to keep (or build) a monument which bears witness to unwanted history.

Cleaning up the Past

In this paper, I have looked at ways of relating to the past through practices of cleaning in relation to monuments. The process of cleaning up history is rarely ever finished. Bits and pieces of destroyed monuments continue to be kept as souvenirs or as museum objects. *Memoryscape*s can be purified using the remainders or former sites of monuments. Artworks may deal with the emptiness that is left behind when monuments are removed or simply never made by representing the void that the lack of a monument creates. The negotiation around what to keep and what to discard reveals the ambiguities immanent to dealing with global entangled histories, and the permanence of memory beyond the material.

Practices that establish cleanliness are practices that create order and establish order-compliance (Bloch et al, 2016:p.161). A changing of orders and values within a society or political system can be introduced by establishing monumental cleanlieness, or cleaning away monuments can be the last step undertaken in the process of political change (see Etzioni, 2004:p.34). More importantly, cleaning up tangible leftovers of monuments does not mean that memories disappear. Rather, interaction with monuments, be it constructive or destructive, is a way of working on memory, of coming to terms and negotiating different perspectives on a society's past.

Cleaning up monuments also sheds light on questions around the ownership of history and of historiography. Dealing with monuments is always an activity that controls the ways in which the past is perceived and though about in the present. In this sense, practices of cleaning up the past are always creative appropriations and can renew a collective identity by means of negotiating and sorting out the collective memory.

---

16 see description on the project website: Mémoires Perdues. (n.d.) *Das Projekt.* [Online] Available from: http://www.memoiresperdues.de/info-memoires/ (Accessed 15th May 2017).

17 "Zu Ehren der für das deutsche Vaterland gefallenen kameruner Soldaten – 1914-1918".

Bibliography

Aday, S., Cluverius, J. & Livingston, S. (2005) As Goes the Statue, So Goes the War. The Emergence of the Victory Frame in Television Coverage of the Iraq War. *Journal of Broadcasting and Electronic Media*, 49 (3), 314-331.

Aldrich, R. (2006) Introduction. Colonies et commémoration. *Outre-mers*, 93 (350), 5-26.

Bloch, A., Bredenbröker, I., Drazin, A. & Knowles, R. (2016) Cleaning Up after Gropius. Eine Ethnografie des Schmutzes. In: Bittner, R. & Krasny, E. (ed.) *Auf Reserve: Haushalten! Historische Modelle und aktuelle Positionen aus dem Bauhaus*. Leipzig, Spector Books.

Delauney, M. (2004) *Au pays des Ravenala. Madagascar 1954-1975*. Nizza, France Europe Éditions.

Dessem, M (2017) How Did We Treat Monuments to White Supremacists When They Weren't Our White Supremacists?. *Slate's Culture Blog*. [Online] 13th August 2017. Available from: https://slate.com/culture/2017/08/read-the-allied-order-to-destroy-nazi-monuments-in-germany.html [Accessed 23rd March 2018].

Douglas, M. (1966) *Purity and Danger. An analysis of concepts of pollution and taboo*. London, Routledge & Kegan Paul.

Etzioni, A. (2004) Holidays and Rituals. Neglected Seedbeds of Virtue. In: Etzioni, A. & Bloom, J. (ed.) *We Are What We Celebrate. Understanding Holidays and Rituals*. New York, New York University Press, pp. 3-40.

Finney, A. (2015) The Cycle of Impermanence: Collective Memory and the Politics of Monuments in Managua, Nicaragua. *Journal of Undergraduate Research*. [Online] 1 (1). Available from: http://laic.columbia.edu/journal-undergraduate-research/the-cycle-impermanence-collective-memory-and-the-politics-monuments-managua-nicaragua/_[Accessed 24th March 2018].

Fisher, M. (2011) The Truth About Iconic 2003 Saddam Statue-Toppling. ProPublica/ New Yorker investigation finds the media got it wrong. The Atlantic. [Online] 3rd January 2011. Available from: https://www.theatlantic.com/international/archive/2011/01/thetruth-about-iconic-2003-saddam-statue-toppling/342802/ [Accessed 24th March 2018].

Francois, E. & Schulze, H. (eds.) (2001) *Deutsche Erinnerungsorte I*. München, Beck.

Gabriel, M.-C. (2017) Quand les frondeurs deviennent des héros nationaux. La renaissance du Monument des héros nationaux à Ouagadougou (Burkina Faso) à l'issue de l'insurrection populaire de 2014. *Cahiers d'Études Africaines*, 227, 691-718.

Galibert, D. (2011) *Les gens du pouvoir à Madagascar. État postcolonial, légitimités et territoire (1956-2002)*. 2nd ed. Paris, Karthala.

Groth, E. (1989) Galgen am Kilimandscharo. Das schmähliche Ende der Karriere eines deutschen Nationalhelden. *Die Zeit*. [Online] 1989 (04). Available from: http://www.zeit.de/1989/04/galgen-am-kilimandscho [Accessed 25th March 2018].

Hahn, H. P. (2016) Der Eigensinn der Dinge. Warum sich Objekte in bestimmten Momenten anders verhalten, als sie es sollten. *helden. heroes. héros*, 4 (1), 9-15.

Hedrick Jr., C. W. (2000) *History and Silence. Purge and Rehabilitation of Memory in Late Antiquity*. Austin, University of Texas Press.

Herz, R. (2009) *Lenin on tour*. Göttingen, Steidl Verlag.

Hinz, B. (1993) Denkmäler. Vom dreifachen Fall ihrer „Aufhebung“. In: Diers, M. (ed.) *Mo(nu)mente. Formen und Funktionen ephemerer Denkmäler.* Artefact 5. Berlin, Akademie-Verlag, pp. 299-312.

Klausmeier, A. (2013) Paesaggio monumentale Muro di Berlino. Ancora un „monumento scomodo“?. In: Franz, B. & Kofler Engl, W. (eds.) *Umstrittene Denkmale. Der Umgang mit dem Erbe der Diktaturen.* Veröffentlichungen des Arbeitskreises Theorie und Lehre der Denkmalpflege e.V. 22. Holzminden, Verlag Jörg Mitzkat, pp. 156-163.

Lentz, C. (2016) A Lasting Memory. The Contested History of the Nkrumah Statue. In: Lundt, B. & Marx, C. (eds.) *Kwame Nkrumah 1909–1972. A Controversial African Visionary*. Historische Mitteilungen. Beiheft 96. Stuttgart, Franz Steiner, pp. 153-184.

Lentz, C. (2017) Ghanaian “Monument Wars“. The Contested History of the Nkrumah Statues. *Cahiers d'Études Africaines,* 227, 551-582.

Macdonald, S. (2009) *Difficult Heritage. Negotiating the Nazi Past in Nuremberg and Beyond.* Abingdon, Oxon, Routledge.

Meier, H.-R. (2013) Umstrittene Denkmale und Denkmäler der Diktaturen. Zur Aktualität der Fragestellung – eine Einleitung. In: Franz, B. & Kofler Engl, W. (eds.) *Umstrittene Denkmale. Der Umgang mit dem Erbe der Diktaturen.* Veröffentlichungen des Arbeitskreises Theorie und Lehre der Denkmalpflege e.V. 22. Holzminden, Verlag Jörg Mitzkat, pp. 16-19.

Montero, C. C. (2010) *Study on how the memory of crimes committed by totalitarian regimes in Europe is dealt with in the Member States.* [Online] Madrid, Institute for Public Goods and Policy. Centre of Human and Social Sciences. Available from: http://digital.csic.es/bitstream/10261/34366/1/Closa_Memory_of_crimes.pdf [Accessed 24th March 2018].

Mühling, J. (2011) Lenins Nase. Der Tagesspiegel. [Online] 20th March 2011. Available from: https://www.tagesspiegel.de/kultur/die-stadt-lenins-nase/3965812.html [Accessed 24th March 2018].

N'Guessan, K. & Späth, M. (2016) In the crossfire of commemoration. Entangled histories of the tirailleurs africainsas heroes in national-day parades in Francophone Africa and France. *Working Papers of the Department of Anthropology and African Studies of the JohannesGutenberg University Mainz* 169. Available from: https://www.blogs.uni-mainz.de/fb07-ifeas/files/2019/07/AP-169.pdf [Accessed 30th July 2019].

Nora, P. (1989) Between Memory and History. Les Lieux de Mémoire. *Representations,* 26, 7-24.

Oberlé, P. (1976) *Tananaríve et l'Imerína. Description historique et touristique.* Antananarivo, Philippe Oberlé.

Pigenet, M. & Tartakowsky, D. (2006) Vues d'Afriques. Entretien avec Jean-Pierre Dozon, Jean Fremigacci et Lucile Rabearimanana Paris, 18 avril 2005. *Sociétés & Représentations,* 22 (2), 81-92.

Ranaivoson, D. (2005) *Madagascar. Dictionnaire des personnalités historiques.* Saint-Maur-des-Fossés, Éditions Sépia; Antananarivo, Éditions Tsipika.

Schmid, H. (2014) Ich bin ich. *Telepolis.* [Online] 16th February 2014. Available from: https://www.heise.de/tp/features/Ich-bin-ich-3362436.html [Accessed 25th March 2018].

Schofield, M. (2015) How Germany dealt with its symbols of hate. *McClatchy.* [Online] 30th July 2015. Available from: http://www.mcclatchydc.com/news/nation-world/world/article29557972.html [Accessed 23rd March 2018].

Späth, M. (2015) *Ein günstiger Augenblick. Das Jubiläum der Unabhängigkeit in Madagaskar*. Mainz, Department of Anthropology and African Studies. Johannes Gutenberg Universität Mainz.

Speitkamp, W. (ed.) (1997) *Denkmalsturz. Zur Konfliktgeschichte politischer Symbolik*. Göttingen, Vandenhoeck & Ruprecht.

Speitkamp, W. (2000) Kolonialherrschaft und Denkmal. Afrikanische und deutsche Erinnerungskultur im Konflikt. In: Martini, W. (ed.) *Architektur und Erinnerung*. Formen der Erinnerung 1. Göttingen, Vandenhoeck & Ruprecht, pp. 165-190.

Targiel, R. (2002) Zum Schicksal Frankfurter Denkmäler nach 1945. *Mitteilungen. Historischer Verein zu Frankfurt (Oder) e.V.*, 2, 35-40.

Zeitz, J. (2017) Why There Are No Nazi Statues in Germany. What the South can learn from postwar Europe. *Politico Magazine*. [Online] 20th August 2017. Available from: https://www.politico.com/magazine/story/2017/08/20/why-there-are-no-nazi-statues-in-germany-215510 [Accessed 23rd March 2018].

Zerubavel, E. (2003) Calendars and History. A Comparative Study of the Social Organisazion of National Memory. In: Olick, J. K. (ed.) *States of Memory. Continuities, Conflicts, and Transformations in National Retrospection*. Durham, Duke University Press, pp. 315-337.

Section Six

Artistic Perspectives

Growths: Contemporary Art between Cleaning and Value

Isabel Bredenbröker

Works that were featured in the *pop up* exhibition which accompanied the workshop in Frankfurt incorporate things that grow outside the realm of our attention / intention: grime in the niches of everyday habitat, by-products of bodily presence, the evolving nature of organic matter and psychological movements at the blind spots of the human mind. Just like a mushroom, another organism that can befall a space and be gone like a ghost, this show invites to think about issues of dirt and cleanliness beyond academia as a laboratory environment. Instead, refuse, seemingly unwanted reactions, animalistic social interaction and material deterioration take the stage. They become the centre of attention and transform the practice of thinking and producing (as a way to make order, to clean up), into it's very opposite. *Wagehe Raufi*s works manifest the tension of having or not having control over an experiment. Different materials, representing different functions, life worlds and modes of existence interact with one another. The artist and the observer can never be sure whether they are looking at a controlled change of matter or whether they are just left as the curious onlooker facing a mutation. *Mia Bencun*'s work is constantly changing and deteriorating. In a slow process, the ephemerality of material caused by reactions between different substances becomes apparent. Glass, the material of modernity, serves as a frame and body. But it is also an accomplice to transformation, hereby betraying the promise of unchanging cleanness it once had. *Andreas Koch* reproduces an everyday object that gets a lot of use but little attention. Rather, one might prefer not to look too closely at what just serves a purpose. The straightened copy of a shower curtain invites the viewer to study its wear-and-tear. *Kerstin Gottschalk* brings back a material that is an easy-to-make DIY favourite with kids and decoration afficinados. She appropriates salt-and-flour based play dough to let a mirror morph, lose its shape and become overgrown. *Chrischa Oswald*'s video explores a mother and daughter's relationship with one another, appropriating animal behaviour. The act of licking each other's faces provokes the question of physical intimacy – a sign of kin relationship, closeness or intimate care with hygienic functions? *Nikolaus Kockel* and *Robert Schittko* perform throughout the workshop and collect rubbish and dirt. These are turned into epoxy resin dishwasher tablets that are then for sale. The process of devaluation in 'making dirt' is hereby reversed, the remainders of academic thought production conserved like a fossil in amber.

in: Bredenbröker, I., Hanzen, C. & Kotzur, F (eds) 2020: *Cleaning and Value. Interdisciplinary Investigations*, Sidestone Press (Leiden), pp. 231-234.

The location provided for the exhibition is in itself a space that imposes rules of cleanliness and proper maintenance. No artwork is allowed to change the setup of the rooms, change the wall surfaces or other. Therefore, the exhibition also imposes itself on the location whilst finding its context in the natural habitat of academic get-togethers.

Growths

9.6.2017
6pm-9pm

Mia Bencun
Kerstin Gottschalk
Andreas Koch
Nikolaus Kockel
Wagehe Raufi
Robert Schittko
Chrischa Oswald

Mother Tongue (2013) Interview

2-Channel-Installation, HD Video, sound, b&w, 4.10 Min. (Loop)

I: Isabel Bredenbröker, C: Chrischa Oswald

I: I would like to ask about working with your family and to consider whether this aspect of your work can somehow be brought in connection with the topic of cleaning, or with value. Especially in relation to the kind of intimacy that Mother Tongue tests out, is this intimacy a value in itself? Maybe as a precondition for making it? Could you also have made this video with someone else? Is what happens here cleaning?

C: Firstly on the question of whether I could have made the video with someone else: no. At first, my mother refused to collaborate at all, but for me her participation was really important. I had this set idea of licking her face, the way cats do it, or how it is performed among animals as part of their body care routine. And I was interested in seeing what would happen if this was transposed into the realm of human social interaction, where today the cleaning of bodies is a completely different thing. So maybe there were certain similar dimensions to bodily care among humans back in the day, like inspecting someone else's scalp for lice. Surely that must have been an intimate moment. Or when mothers bathe their children. But in relation to touch, a direct kind of physical interaction with our tongue is more likely to be read with a sexual connotation and is something very intimate. That's why it can create such tension, it produces a kind of conflict between the pure act and the relationship between mother and daughter, who in the video appear to be quite innocent, dressed in white. I've always termed the beginning and the end of the video as the "NIVEA moment", the image of mother and daughter in an intimate portrait setting.

I: Why NIVEA?

C: I am always playing with the idea of the idyllic family and intimate togetherness. I've seen this moment shared between mother and daughter in some beauty advertisement before, this act of taking care of each other's bodies. Even if that only means that one person leaves a bit of lotion on the nose of another. But transporting this idea of: that's good for you, that's what care is, that's what gives you a spotless complexion or somehow nurtures you. I wanted to record a moment that worked well for advertisement purposes, an idyllic situation, and then go on to break this moment up. Which, in the end, is still a very pure gesture for me. When cats lick, it's cute and everyone is taken by it. But as soon as one shifts this gesture to another context, things like incest and so on come into mind. That was my mother's main

in: Bredenbröker, I., Hanzen, C. & Kotzur, F (eds) 2020: *Cleaning and Value. Interdisciplinary Investigations*, Sidestone Press (Leiden), pp. 235-238.

concern, the reason why it took me so long to win her over for this experiment. She thought, when this becomes public and people from the town see it, they are going to ask themselves if she is mad, or if I am, or if we both are. And what is wrong with our relationship? This is where I believe the value of this video is somehow not fully appreciated. I mean the value of art. The value of breaking conventions and looking at them from a different angle.

I: I am living with a cat at the moment. Of course, it's just one cat but it kind of looks as if she can take care of cleaning herself very well without the need of another cat.

C: I guess I'm thinking more of mothers and children, cats and kittens. This is why the chance to experience this myself seemed so important to me. How does it feel? What is it like? And because I work a lot with my family, it was not an option to use someone else who would perform as a mother for me.

I: Which means that family in itself serves as a kind of value in your work; a level of mutual respect and trust. Does that then mean that you could not film these transgressive scenes, scenes somewhat uncharacteristic as far as human behaviour goes, with other people? Or that you would find it less productive?

C: I think I could. But I'm really more interested in the dynamics that unfold within the family. You can also read it as a kind of therapy, if you want: I am trying to understand certain structures. Whilst doing so, I am testing the boundaries. How far is my family willing to play along? This is where the value of trust is really important. I could pay other people to do this and that's not a problem at all. But I also didn't want it to look too much like it was all just an act. There had to be an element of the real and real relationships in it at all times so that everything that resonates within our family has the chance to resonate in the works.

I: That means this significance of 'having a family' and of sharing, of being a common unit, is really important for these dynamics.

C: Yes, I think so. But I'd like to come back to the value of art once again. That's an important point for me because I also had to make it clear to my family who are not very interested in art. To explain to them why it holds such a high value for me and why I want to spend my life doing art. Why I invest in work and thoughts. I want to show them what this can mean and what a different kind of world or experience it can create. But of course, the value of the family and of how I am able to work with them, trust and everything my family gives to me which may not exist in other families, the fact that I am able to use this as a material for my work is of course very important.

I: It becomes something different if you end up paying someone else for the part. The contract is different and emotions are different, the experience will be a different one. And what kind of effect does this setup then have on you in the end? Was it cleansing in any way?

C: The act itself was somehow disgusting for both of us. Saliva starts smelling when it dries in contact with air. My imagination of how it would be was a lot more romantic than how it turned out. I might be just imagining this, but perhaps it had a somewhat cathartic effect on my mum both during the shooting but also afterwards when the work was featured in exhibitions.

I: Yes, she got over her fear.

C: Exactly. And I was quite amused when I heard how different people reacted to the work. One of my best and oldest friends, who also knows my mum and likes her a lot, she basically could not look at it at all. Another person told me he thought it was very romantic. I had expected to represent an expert in the video. I thought that could be a powerful image. So having achieved that is already an achievement, even though I found the physical interaction repulsive.

I: You meet as two adults in the video. If we go back to images from the animal kingdom, bodily care is more likely to occur between mothers and their young children. Like suckling or breastfeeding for humans. The mother takes over a task which the child is still incapable of doing themselves. But you are able to wash your own face these days...

C: But I can't lick my own face.

I: Yes, you do need someone else for that.

C: No, but you're right and that was also the point. Now that I am grown up it comes across even more weird. I am a sexualised body and don't have the purity of a child anymore.

I: Adult bodies do not necessarily have to be sexualised.

C: Yes, but charged. It's clear that I am sexually mature, I know desire and lust. For the child, these things do not resonate in physical interaction. I am aware of that, even when trying to be as neutral as possible and to just let the interaction stand for itself. But for me this tension is always present. So my play includes a different dimension.

I: That means that the act of a body being cleaned by another body, involving the direct interaction with bodily fluids which may already have a sexual connotation, is likely to be interpreted as purely functional – an act of cleaning – when the body that is being cleaned can pass as morally pure. It seems almost paradoxical that one should only be allowed to bodily clean that which is already considered somewhat pure. What does it mean if that purity is not a given? That the persons who are involved in this act are polluting one another?

C: Yes, it may be a kind of moral pollution I would say. That was exactly my mother's fear even if she knows that there is nothing sexual between us but based purely based on what a viewer sees may think.

I: Unorthodox cleaning. An established method in a wrong or new context. In this case also with new actors. I mean, the child is also active in the cleaning and is not just being cleaned.

C: This mutuality was important in order to see whether it changes anything to change roles or to have a fixed role.

I: Did you rehearse?

C: Not really. After all the convincing...

I: How long did it take to win her over?

C: It took about three months until she gave in. And then some more time until we actually filmed it. The idea was to do it and then see whether I would be allowed to show it but in the end it was kind of spontaneous and unannounced. She suddenly said, "So do you still want to do this video thing? Because we're going to see your grandmother later and I have a bit of time before we leave." And I thought, "great, now I'm actually not prepared at all." The only white wall was in my parents' bedroom, so we shot it there. We did several takes. Of course, there was plenty of laughing in between and then we would start over again.

I: But the video that was chosen in the end seems to be basically uncut.

C: Almost, yes. But then she said she didn't like the way she looked on screen. In the end we gave it another go two months later.

I: So once again with feeling?

C: Well, what would feeling be here? She tried to approach it in a more relaxed manner. I set us the goal to do it uninterrupted for a certain time so that I would not have to cut. But in the end everything went really fast. The most time-consuming thing was the convincing and talking.

I: So the version that you selected in the end is from the second take?

C: Yes. That was when she agreed that it would be okay for her to show it. But she wanted to have a say in what the final video would look like. Nobody operated the camera, we just put it there and turned it on.

I: That is very intimate, without even having a third person in the room.

C: Yes, only the two of us in my parents' bedroom.

I: Did it change something in your relationship to your mother? Or did you expect it to do something?

C: Not really. I think instead it showed me that my mother really has to love me in order to do something like this for me. If one is truly afraid of being judged, then it takes a serious effort to overcome this fear. She was torn between her intention to support me and the values of the society around her which collided with my ideas. It showed that she supports and acknowledges my art practice and wants to see her daughter happy.

I: Which in itself is also a way of showing value.

Figure 1-3. Stills from Mother Tongue (Photo: Chrischa Oswald).

Mehl, Salz und Wasser zu einem Teig verarbeitet und auf die Spiegeloberfläche aufgetragen (2017)

Flour, salt and water, prepared as a dough and applied to the mirror surface, size varies (2017)

Kerstin Gottschalk

Above and right: Bathroom before installation.
Next pages show the work installed in the bathroom.

(Photos: Kerstin Gottschalk)

ww.creo.com

Duschvorhang (2005/2015)

C print, pegs, aluminum

Andreas Koch

Above and right: Shower Curtain .

(Photos: Andreas Koch)

Figure 1. Tools used by the artists during the Cleaning and Value Workshop in Frankfurt, 2017. (Photo: Research Training Group 'Value and Equivalence').

Figure 2. Artists at work during the Cleaning and Value Workshop in Frankfurt, 2017 (Photo: Research Training Group 'Value and Equivalence').

Cleaning and Value

Robert Schittko, Nikolaus Kockel

Cleaning is the process of separating value from dirt. As such, it is central to human existence and the foundation of the man-made environment in which we live today. For decades, modern and postmodern art has been interested in the margins of this environment, where safety and sanity succumb to chaos that is dirt. This interest was deepened in the art history of the twentieth century, cumulating in dust sculptures and pictures of excrements. In this regard, the work of some artists could be seen as complementary to cleaning, as systematic filth.

From a social point of view, art proves to be an omnivore who, by means of her power of discourse and the speculative nature of the art market, creates value from what has been discarded. She shows herself to be a true, a better alchemy that does not need lead to create gold (value), but is satisfied with dirt.

We want to ironically relate to this value chain in our project: Disguised as cleaners, we seize the dirt that accumulates during the first day of the workshop and pour it into epoxy (fig. 1-2). Just as a memorial diamond transcends simple carbon of cremation to become an eternally lasting, sparkling memento of a loved one, we will preserve the ephemeron, the leftovers and waste of the scientific debate in crystal-clear synthetic resin. These pieces of jewellery can then be purchased during the exhibition and thus be upcycled as objects of value (fig. 3).

in: Bredenbröker, I., Hanzen, C. & Kotzur, F (eds) 2020: *Cleaning and Value. Interdisciplinary Investigations*, Sidestone Press (Leiden), pp. 249-250.

Figure 3. Sales desk with the produced diamonds, taken during the soiree on the first workshop day (Photo: Research Training Group 'Value and Equivalence').

In between contemporary stones (2016)

LED panel, steel structure, agar agar, retention granules, ink, pigments, multiple sculptures

Wagehe Raufi

Above: In between contemporary stones.

(Photos: Johannes Lenzgeiger)

Contemporary Art and the Gaze of an Archaeologist: An Interpretative Attempt of Decay and Lost Evidence

Anna Langgartner

The gaze of an archaeologist is normally applied to old and badly preserved objects. They are mostly fragmented and incomplete: metallic parts irreversibly changed due to processes of oxidation, or organic parts forever lost due to decomposition over the course of hundreds of years. Thus, the gaze stands in a persistent dialogue with the object of its interest, wherein every new question is rewarded with new information. The final aim of this dialogue lies within the augmentation of knowledge about lost cultures from the past.

Looking at the objects of Wagehe Raufi's work "In Between Contemporary Stones" the archaeologist's gaze is confronted with a conglomeration of various organic and synthetic materials seemingly mashed together into clusters of diverse colour, texture and form. Therefore, the first question for an archaeologist would be: What exactly are these objects made of? Visible to the eye are pieces of a foam-like material, plastic film and colour pigments such as green, pink and yellow. A more elaborate analysis would reveal remainders of make-up, sanitary products, different kinds of plastic, corn starch and other traces of aliment, and finally hydrogel and agar-agar. These last two materials – better described as water storage granules and gelatine extracted from algae – form a kind of matrix when they are mixed with water, which holds these vastly different materials together.

Examining the materials independently already allows for conclusions about the use of products that occur in everyday life, the consumption of food, rites of personal hygiene, or technological knowledge of extracting and crafting organic and synthetic materials. A classical archaeologist may draw comparisons to the excavation of roman sewage systems, or garbage dumps within exhausted water wells, for example of a Hellenistic Greek polis. Especially organic materials are of great interest in these contexts for they are often decomposed and can only be identified through costly methods of analysis.

Equally important throughout an archaeological analysis is the context and function of the objects. The clusters of Wagehe Raufi's artworks are crafted objects and not just a coincidental accumulation of different substances such as would be found in the remainders of garbage dumps or sewage systems. Time and effort were needed for

in: Bredenbröker, I., Hanzen, C. & Kotzur, F (eds) 2020: *Cleaning and Value. Interdisciplinary Investigations*, Sidestone Press (Leiden), pp. 255-256.

their creation, and therefore, an intentional purpose or function can be expected. So, what are these objects made for?

An archaeologist relies on the information gained from knowledge of the excavation context, the location where objects are found, and written sources to identify an object's function. For example, an anchor identified by size, form and material, found in a sanctuary with a dedication inscription, would have served as a votive offering for a safe passage through the Mediterranean Sea. The same anchor, found within a shipwreck on the bottom of the sea, would have made its original function as an anchor explicit.

Here, we are dealing with a work of art produced for a contemporary audience that has been presented in various exhibitions. The context of an art exhibition and signs with titles, year and material, which frame the objects as artworks, make their function as art more than evident.

An interpretation of the work in an archaeological sense could be, first of all, described as an insight into the artist's life – solely as a representative of her culture and time – through the combination of all the different materials and objects surrounding her on an everyday basis. In comparison with artwork of the same time and culture, Wagehe Raufi's works reveal an outright experimental character, which defines the term 'sculpture' in an innovative and uncharacteristic manner.

But the questions remain: what does the artist want to transport and what does the viewer receive? Should the clusters be understood as a critical comment about material usage and waste? Or does the artist aim to evoke basic feelings inside her viewers head, such as joy, fear or disgust? Such ideas need to be verified by written sources, for example interviews with the artist or viewers, and by comparison with similar artworks. Only with a profound basis of evidence could an archaeologist make a prediction or draw a conclusion about further interpretations, because most of the time he is dealing with objects crafted by persons of long gone cultures. These persons received a different education, had a different use of media and left a material culture different from that of our modern European culture. Therefore, the archaeologist needs evidence to proof his ideas and cannot draw from his own cultural experience like a hypothetical viewer of Wagehe Raufi's work, who he tries to reconstruct.

But one must keep in mind, that contemporary art in general possesses a highly ambiguous character. There is rarely only one intended interpretation of an art object, or only one feeling the artist likes to evoke in its viewers head. To a certain degree this ambiguity is transferable to artworks of lost cultures, for example to a classical Greek vase painting. Hence, the archaeologist's efforts to reconstruct the presumably correct interpretation of ancient art works will never lead to a complete understanding of them, but it will come close.

And in a few hundred years from now on only the plastic parts of the Wagehe Raufi's objects will remain and the original function as art will be eliminated without coincidentally passed on written sources. A future archaeologist would not be able to identify the objects as artwork.

Bruise 13 (2017)

Glass, latex, copper, aluminium, 100 x 70 cm

Mia Bencun

The latex pieces react with the copper wire and develop a brown and green patina at the points where the materials touch. The copper leaves traces of its presence in the susceptible latex material. The objects are fragile and of limited durability. Each work expires at different intervals and in its own way.

In the cultures of one of my home countries, the cycles of time, such as reincarnation and the eternal recurrence, are considered more important than its linear course. There is no focus on the irretrievable or on accumulation, but on the regularity of renewal. Transitoriness is nothing that one has to put up with due to technical inadequacies, but it experiences a cultural upgrading – in a certain way, one decides in its favour. Objects are also born and die, just like animals and humans. For this reason, I do not prefer the particularly durable materials, which are suitable for buildings and monumental works of art, but ephemeral and living, so to speak.

Copper and UV rays damage the sensitive latex cast. Gradually, it begins to change uncontrollably, tearing and darkening. The cycle of decay requires a renewal of the works through repair castings. Conditionally, decay can be delayed with the help of silicone oil, but it cannot be prevented. Repeated restoration results in new forms and compounds, which in turn produce residues. A moment in which the real is catching up with us. A flow between the permanence and impermanence of the body. Residues refer to the transient and the inevitable changes that time brings.

Humans avoid confrontation with these residues, since they are based on bodily residues as well as socially repressed residues. However, these residues are a constant thread, which humans cannot deny. Fragile and degraded, the decaying remains have lost their original form and are essentially fragmented. They speak of a past function or former state of wholeness, but were torn loose and are drifting to the periphery. Unlike garbage or thrash, which are permanently displaced, the edges remain floating and elastic.

in: Bredenbröker, I., Hanzen, C. & Kotzur, F (eds) 2020: *Cleaning and Value. Interdisciplinary Investigations*, Sidestone Press (Leiden), pp. 257-260.

Left: Bruise 1.
Above: Bruise2.

Photos: Mia Bencun